# Fifty ways to stuff the planet

Little things that you can do to help
global warming on its way

with

# Dave Blunt and Phill Knight

If we don't mess it up,
somebody else will.

First published in 2008

Copyright © 2008 by Andy Sharp

ISBN 978-1-4092-0520-3

# Blunt speaking

Stuff it, I mean, all this stuff about global warming, you'd have to be a bit of a spanner to think we can stop what we've started. Who's going to give up their motor, big screen telly or weekends getting pissed in Barcelona just so that some other arsehole can do it instead. Get real, we'll keep on burning coal, oil and gas until there isn't any left.

Besides, I mean, we've been on this planet for about a million years and we're just getting to the exciting bit, the tipping point, the point of no return, the time the shit hits the fan and I want to be around when it happens.

How fast will the sea level rise? When will malaria get to Paris? Which small town in the midlands will lose half its roofs in an unscheduled tornado? When will I be able to navigate the North West Passage in a pedallo? How big a splash will there be when the West Antarctic Ice Shelf slips into the Southern Ocean?

So let's give climate change a good kick up the arse, let's piss off the greenies and see how well we can stuff things up in our own lifetime.

Now, it's a damn sight easier to go with the flow than to swim against the tide. The greenies are right when they say we can all make a difference. Yes, we can. We can all do our bit to speed things up.

This little book shows that you can help climate change on its way simply by carrying on doing ordinary everyday things.

But to do this properly, we might as well do it scientifically. This is the age of bloody information after all. So for each of our tips Phill, sad geek that he is, and myself have kindly worked out how much it costs (*What's the Damage?*), how much effort it takes (*Can I be arsed?*) and how annoying it could be (*Wind-up-ability*)

Charlie D showed us that nature responds to a challenge, survival of the fittest and all that. Let's give it a big one.

## Something of the Knight

Hiya

I'm Phill, Dave's sad scientific mate. I simply reckon that if you understand what you're doing you stand more chance of doing it properly. To stuff things up most effectively this means getting to grips with a bit of basic science.

My old granddad was a self taught engineer. He thought that you could use trial and error to solve any problem. Up to a point, the start of the industrial revolution, he was just about right. From then on in he was well wrong.

For example, the first steam engines worked, but to be honest they were a bit crap. At best you only got about 2% of the energy out that you put in. Working out how to make them better gave rise to the science of thermodynamics. Modern, scientifically designed, power stations do at least 20 times as well.

It was only when I asked my granddad if he could make a telly just by trial and error, without understanding electricity and stuff, that he grudgingly gave in.

So, pay a bit of attention and you too can get to use science for your own ends; however dastardly they might be.

**n.b.** I helped Dave with the sums but have to admit that some of the numbers are now a bit out of date.

But, because we've shown how the sums have been done, it shouldn't be too much trouble to bring them up to date all by yourself.

+ You'll notice there aren't many pictures. This is because Dave and I are crap at drawing. But, we've left spaces so that you can draw your own.

If anyone would like to help us make an illustrated version of the book just go to www.stufftheplanet.com and get in touch.

# Contents

## Around the house

## Magic motors

## You are what you eat

## Spend, spend, spend

## Chuck it out?

29) Chuck out your old telly
30) Throw your tinnies in the bin.
31) Don't throw out those old light bulbs (yet)
32) Keep that old fridge off the scrap-heap
33) Dump that bike, why ride when you can drive.
34) Buy a bike but never use it.
35) Buy a bike for your car.
36) Leave that loft alone.
37) Never wear more than just a T shirt.
38) Play golf in a warm climate.

## S.I.Y. (stuff it yourself)

39) Leave hand tools to the handyman.
40) Have a strimtastic afternoon in the garden.
41) Rake space for a leaf blower.
42) Put down those clippers and get a hedge trimmer.

## Use the system

43) Get into debt
44) Invest unethically?
45) Become a slum landlord.
46) Drain a bog.
47) Fight for your right to parking.
48) Vote for the cynical greedy so and so party

## Seeming to do the right thing

49) Run your motor on bio-fuel
50) Get a wind turbine but forget to insulate the loft

# Stuffing things up around the house

They say charity begins at home. "Look after number one", that's what I say, "no-one else will."

## One Leave a 100W light bulb on all year

"Please switch off when not in use", my arse. What could be easier than forgetting there's an off switch. Not only do you get to give climate change a good kick up the backside, you also get to carry on behaving like an oblivious teenager. Excellent.

*What's the damage?*

There are 8760 hours in a year, so each year a 100W bulb uses 876kWh of electricity. (remember, there are 1000W in 1kW and 1000Wh in 1kWh) Each ordinary bulb lasts about 1000 hours you'll need about 8 of them.

At 30p for each bulb and 10p per kWh, the total cost is about £90 (8 x 30p + 876 x 10p)

Making 1kWh of electricity in the UK produces about 0.43kg of $CO_2$

So, you'll emit about 377 kg of $CO_2$ (876 x 0.43) at a cost of about 24p/kg. (pence per kilogram)_

*Can I be arsed?*

Bit of an arse changing the bulbs, but going all dark should give you a clue when it needs doing.

*The Wind-up-ability*

You may get some domestic grief for this, so play it by ear. Unlikely to piss off anyone else 'cos they won't know about it._

## **Two** Never, ever, switch off your PC

*What's the damage?*

A typical PC uses about 175W of electricity. Switch on the power saving features and this can drop as low as 35W. So, to maximise your PC's stuff-up-potential you'll have to go into settings and disable all the planet saving gubbins.

Simply owning a PC costs about £120 per year (£600 spread over about 5 years) so it probably isn't worth getting one specially.

Otherwise, each year it uses about 1533kWh of energy (175W x 8760 hours), which costs about £153 (1533 x 10p) and produces about 659kg of $CO_2$ (1533 x 0.43).

This works out at 41p for each kg of $CO_2$, (if you include the cost of the PC), or 24p/kg (if you don't.) (i.e. (£(153 + 120)/659kg)

+ there's not only the energy used when the PC's working, however pointlessly, there's also the energy that got used to make the thing in the first place.

*Can I be arsed?*

Piece of piss

*The Wind-up-ability*

Absolute bobbins unless you happen to live with a yoghurt weaver. No one else will ever notice.

## Three Leave the fridge door open

A fridge motor only kicks in when the inside of the fridge gets too warm.

You can make this happen more often by turning down the thermostat, by blocking the airflow around the black pipes at the back (where the heat from inside the fridge gets dumped) or by leaving the door open. It all depends on whether you want the stuff inside the fridge to stay cold.

*What's the damage?*

A typical fridge motor runs at about 150W. So, if you could get it to stay on all year it'd use 1314 kWh of electrical energy (150W x 8760hours), cost about £131 (1314 x 10p) and release about 565kg of $CO_2$.

Because you definitely need a fridge, what with your lagers and all that, it wouldn't be right to include the cost of buying the fridge in the first place. Hence, the net cost is about 24p/kg of $CO_2$.

*Can I be arsed?*

Shutting the fridge door can become a bit of a habit, but you could always use a couple of bits of old chuddy to stop it closing properly. That's if you don't mind missing a "fresh white blob on the pavement opportunity".

*The Wind-up-ability*

The constant hum could well put your housemates in a radge, so be careful.

## Four Leave the tap running (don't fix that drip)

You might think that water just comes out of the tap, but before it gets there it has to be treated and pumped. Both these things use energy.

Put an empty 1 litre bottle under a dripping tap and time how long it takes to fill up. You could then work out how much it'd drip in a day, a week or a year.

*What's the damage?*

Suppose it took 10 minutes to fill the bottle. That would be 6 litres an hour, 144 litres a day (6 x 24) or 52,560 litres a year (144 x 365)

Treating and pumping 1 cubic metre of water (1000 litres) is estimated to emit 0.404 kg of $CO_2$. So the total $CO_2$ emitted here, by our 52,560 litres, would be about 21kg.

If you've got a water meter, you'll find it costs about £1.75 per cubic metre. So the total cost would be about £92 (£1.75 x 52.56) and the cost per kg of $CO_2$ would be £4.38 (£92 / 21kg).

i.e. a bit stuffing pricey, unless, of course, you don't have a meter.

*Can I be arsed?*

If you do have a meter then, against your better instincts, you might find yourself getting the drip fixed and saving the coin for something a bit more cost effective.

*The Wind-up-ability*

Dripping taps can be really annoying. The Chinese water torture is no joke and if it's a cold tap dripping into your hot bath you might even piss off yourself.

## Five Open all the windows and turn up the thermostat

A regular central heating system runs at about 15kW, but it only does it when the thermostat tells it to.

Turn the thermostat right up, open the windows if it gets too hot, and let your old boiler realise its full stuff-up-potential

*What's the damage?*

For each kWh, gas costs about 4p and emits 0.19 kg of $CO_2$.

So, running a 15kW gas boiler for a year will use 131,400kWh of energy (15kW x 24hours x 365days), cost about £5256 (131,400kWh x 4p) and emit 24,966kg of $CO_2$ (131,400kWh x 0.19kg) This gives a cost of about 21p/kg of $CO_2$

Fuel oil releases 2.68kg of $CO_2$ per litre so, if you got through a full 900litre tank it'd emit 2412kg of $CO_2$. The current price of oil is about 36p per litre so the full tank would cost £324. This gives a cost of about 13p/kg of $CO_2$

Household coal costs about £200 per tonne (1000kg) and each tonne emits about 2420kg of $CO_2$. This gives a cost about 8p/kg of $CO_2$.

Not only is burning coal the cheapest, it's also by far the most polluting. As the miners used to say, before Maggie T showed 'em what's what, "Coal is king". At least it is when it comes to stuffing the planet.

*Can I be arsed?*

A doddle, but I'm not sure the boiler's going to like it and, if you're using coal or oil, you'll have to have to put up with endless deliveries.

*The Wind-up-ability*

If you burn gas or oil, you're only likely to wind up the hard core yoghurt weavers, no-one else will notice. A good plume of coal smoke, on the other hand, can piss off an entire neighbourhood.

## **Six** Rediscover the joys of a real fire.

You don't need central heating to burn stuff, you can do it with an open fire. So open up that old fireplace and get polluting.

*What's the damage?*

As we've just seen, by burning coal you can emit a kg of $CO_2$ for just 8p ($1/3^{rd}$ of the price of doing it with electricity). Even if the risk of getting bubbled means you have to use smokeless fuel, it's still comes out at only 10p/kg.

*Can I be arsed?*

It's a bloody good job that coal fires look nice, 'cos they're a pain in the proverbial to keep going. You don't just have to keep humping coal you also have to keep cleaning it out. Maybe you could just set fire to a big heap of coal outdoors....

*The Wind-up-ability*

In the Great Smog of 1952 over 4000 Londoners died from the effects of smoke pollution from open fires.

Not only do the little smoke particles clog up your lungs but the fumes contain dilute sulphuric acid.

A good smoky fire can cause coughing fits right across your own neighbourhood as well as adding to acid rain everywhere else.

## **Seven** Make every day a washing day

The amount of energy a washing machine uses depends on its energy rating and how hot you do the wash. For a full cycle, an A rated machine on a low temp wash uses as little as 0.56kWh of electricity. Running a D rated machine on a hot wash can bump this up to more respectable 1.98kWh.

A tumble dryer uses about 2.5kWh per load, but only has about half the capacity of a washing machine so you have to use it twice as often.

*What's the damage?*

One load of washing, and two lots of tumbling, could use up to 7kWh of electricity.

Doing this every day for a year would use about 2555kWh (7kWh x 365days), cost about £256 (2555kWh x 10p), and emit about 1100kg of $CO_2$.

So, the cost is about 23p/kg of $CO_2$

*Can I be arsed?*

That's an awful lot of bloody washing. Best get your mum to do it.

*The Wind-up-ability*

Rumble, rumble, whir, whir, spin, spin. Some people like that sort of racket.

# **Eight** Never switch off at the mains

What the stuff do greenies think the standby button's for. It's for stuffing standing by. If God hadn't meant us to use the remote he wouldn't have given us fingers.

Leaving everything on standby is the opposite of firtling (appearing to be busy while actually doing nothing). You look like you're doing nothing but actually you're hard at work stuffing up the planet.

For example, on standby a typical microwave uses 3W, a typical set top box 10W and a typical video 6W.

Add up all your chargers and gizmos and you could easily be using 100W by doing absolutely diddly-squit.

*What's the Damage?*

Over the course of a year you could easily use 876 kWh of electricity and emit 380kg of $CO_2$.

As we've seen before, for electrical stuff this'll cost about 23p/kg.

*Can I be arsed?*

Who could be arsed not to

*The Wind-up-ability*

Nobody notices this happening now so you're not likely to get anyone's goat.

Indeed, nobody in the UK notices to such an extent that stereos on standby emit 1,600,000 tonnes of $CO_2$, videos 900,000 tonnes, TVs 480,000 tonnes, consoles 390,000 tonnes, DVD players 100,000 tonnes and set top boxes 60,000 tonnes. Keep up the good work.

## Nine Don't get tempted by rechargeable batteries.

Making a battery uses up to 100 times as much energy as it actually stores. Don't waste this advantage by using rechargeables.

*What's the damage?*

A typical carbon zinc AA battery costs about 25p and can supply about 1.4Wh of electricity. Compared to mains electricity this gives a cost of about £179 per kWh (25p/0.0014kWh)

Assuming 50x as much energy is used making the battery as it stores, the overall energy used in the battery's short life is about 70Wh (50 x 1.4Wh) and, assuming this is electrical energy, the overall $CO_2$ emissions would be about 0.03kg (0.07 x 0.43).

This gives a cost per kg of $CO_2$ of about £8.30 (25p/0.03kg)

*Can I be arsed?*

If it weren't for all the extra pollution that comes from making the battery, and then disposing of it, it would hardly seem worth it.

But, don't let these numbers put you off. Even if you did get rechargeables you'd probably bugger them up by forgetting to recharge them properly.

*Wind-up-ability*

Your local recycling coordinator might get a bit pissed off wondering how to keep all these dead batteries out of the local tip, but no-one else will give a monkey's.

# Magic Motors

## Blunt Speaking

Stuff it, I mean, aren't motors brilliant. So big, so shiny, so powerful that even the lardiest arse in the world can feel like a babe magnet and even the most obnoxious little turd can trail a plume of envy down the High Street.

And another thing, have you ever thought about motors and evolution. Charlie D showed us that all that matters is passing on your genes. It used to be that the babes went for the guys who were physically fittest, the best able to catch the food to put on the table. Now it's the dude with the flashiest motor. If he can afford to spend all that money on a car then he can bloody well afford to look after me and the kids.

So, as well as giving lardies a fairer chance in the mating game, motors also help weed out those who aren't really fit for life in the modern world. For example, those that are a bit deaf so they don't hear the motor coming before its too late or those that are prone to asthma who erupt into wheezing fits at the mere sight of a passing truck.

## Ten Leave it running

Seeing as your modern motor spends as much time just hanging around as it does actually getting anywhere, why waste an opportunity to pollute just because you're .

*What's the damage?*

A typical engine at tick-over uses about 1 litre of fuel an hour. A big one will use more, a little one less.

Petrol costs about 90p per litre and burning each litre emits 2.31kg of $CO_2$

So, each hour you leave it running you'll emit about 2.3kg of $CO_2$ at a cost of 39p per kg.

*Can I be arsed?*

Can you be arsed not to?  After all, it can be a bit of a stretch reaching all that way to the ignition key, especially if you've got a good man belly in the way.

Besides, you wouldn't want to get chilly hanging around outside the school gates waiting for the sprogs and you'll be needing all the spare power you can get for your 250W speakers. You wouldn't want to risk a flat battery, would you?

*Wind-up-ability*

Believe it or not, some people find the sound of an idling motor a wee bit annoying. They're the same ones that are likely to complain about the stuff that comes out of the exhaust.

Don't let this put you off. If God hadn't meant us to expose little children to fumes he wouldn't have put exhausts at buggy height.  Encourages them to grow up nice and tall, that's what I say.

## Eleven Make mine a big one

The bigger the motor, the more energy it takes to haul it about, the more fuel it burns and the more greenhouse gases it gives off.

*What's the damage?*

In official tests, a big shiny Range Rover emits nearly 400g of $CO_2$ for each km it travels. But, if you drive it around town its fuel consumption can drop to 12 m.p.g. Converting miles to km, gallons to litres and noting that burning a litre of petrol produces 2.31kg of $CO_2$, this corresponds to a more respectable 543g/km.

Even better, if you only use it to go short distances, the engine doesn't get properly warmed up, it uses even more fuel and, because the catalytic converter doesn't get going until it's hot, chucks out even more pollution.

The only trouble is that while running a motor has never really been cheaper, it's still more expensive than you might imagine. The AA puts the overall cost of a big new 4x4 at just over £1 per mile (assuming 10,000 miles per year). This is 63p per km. So the cost of emitting 1kg of $CO_2$ is a hefty £1.57. (£0.63/0.4kg)

*Can I be arsed?*

So big, so shiny, so high. If you've got the coin, it'd be a tragic waste not to flaunt it.

*Wind-up-ability*

The only thing that gets the goat of a yoghurt weaver more than the sight of a big 4x4 in the city, is a big 4x4 churning up a bridle-path in the countryside.

Make mine a Hummer...

## Twelve Even a crap one can do the job

Even a Toyota Bloody Pious can do the job if you're prepared to drive it far enough. The idea that any motor can be thought of as green is so laughable that only our friends in marketing would dare to suggest it.

Likewise, pick up a cheap motor off the street or at an auction and you can do your bit without it costing **you** the earth.

*What's the damage?*

A £200 motor will still cost you £150 to tax and £250 to insure. If it does 25 mpg, you'll need 400 gallons to go 10,000 miles. This'll cost about £1640 (90p per litre, 4.546 litres per gallon) and emit about 4200kg of $CO_2$.

This gives a total cost of 53p/kg of $CO_2$

At the end of the year, or whenever the MOT becomes due, you can just dump it and get another one.

*Can I be arsed?*

Not only do you get to do your bit for climate change you also avoid having to mix with the old biddies and school kids on the bus. Plus you get to cut up cyclists and park on double yellow lines or even the pavement. You wouldn't want to repress your inner road rage would you? Who knows where that might lead.?

*Wind-up-ability*

A smoky old motor can be a very cost effective way of pissing off greenies. You might consider enhancing the effect with a wide barrelled exhaust and a quality sound system.

## Thirteen Give it some welly.

Quite a few motors these days have a little gadget which shows how much fuel you're using. Take it gently and the little so and so barely moves off "bugger-all".

Back in the olden days there was an event called the Mobil Economy Run, in which tight so and sos would compete to see how little petrol they could use as they toured around the country. The main tricks were to accelerate gently, change up early and, by leaving a good gap in front, go easy on the brakes.

Bearing all this in mind, all you have to do is the opposite. Brake at the last minute and the storm away from the lights. With a bit of practice you'll see your fuel economy plummet and have even more excuse to pop out and fill her up.

*What's the damage?*

With a bit of attention, you should easily manage to use half as much fuel again. It'll cost a bit more at the pumps, and you'll probably have to get your brakes done more often, but the sheer fun of driving like a real man should make it all worthwhile.

*Can I be arsed?*

Of course you can. What fun is a motor if you can't put your foot down. What's the point of that wide bore exhaust if you don't make it roar. What use are tyres if you can't make them squeal. Is there a better smell than burning rubber?

*Wind-up-ability.*

Noise, smell and intimidation make a winning combination. You can frighten the meek, offend the smug and impress the pull, all in one go . Brilliant.

# Fourteen The shorter the better

Just because the newsagents is only 50 yards around the corner, doesn't mean you should get tempted to walk.

Not only does a cold engine use a lot more fuel, that's what the choke's for, but it generally chucks far more crap straight out of the back. In addition, those kill joy catalytic converters don't start working properly until they're really hot and that doesn't happen until you've gone at least a couple of miles.

Remember, if you never leave home without your car keys then it'll just become a habit to use the car for every journey, no matter how short.

*What's the damage?*

A cold engine uses about 30% more fuel and emits about 60% more pollution than one that's fully warmed up. It takes at least three miles for these beneficial effects to wear off, so don't waste them.

*Can I be arsed?*

If you really can't be bothered to use the car for tiddly journeys, just start it up and leave it sat there until you get back. It'll do the job just as well by itself.

*Wind-up-ability*

Leaving the house, getting in the car and then coming back two minutes later will definitely wind up the tree hugging curtain twitcher next door.

In addition, the little cloud of un-burnt fuel will literally get up the noses of pompous pedestrians and the extra car in the traffic jam will slow down everyone else and increase their emissions too.

All in all, a productive exercise.

# You are what you eat?

## Blunt Speaking

Of all the things my dog can do, I sometimes think that the most wonderful is her ability to turn dog food into dog. All those tins of smelly blobs, all those tasteless biscuits, all the sausage rolls dropped by little kids in the precinct. She turns all that trash into beautiful dog. Yet another one of Mother Nature's everyday miracles.

Not that She cares. She just lets us get on with it. As long as we can produce enough reasonably good copies of ourselves, ones that can still turn a Big Mac into something approximating to a human being, then we'll survive long enough to breed and long enough to pass our genes onto the next generation.

Once upon a time, we had to get by with what we could grow, gather or catch in our immediate neighbourhood. If you didn't eat when food was available you'd starve and dead people have a hard time passing on their genes. So, even though lots of us are living in times of plenty, we still stuff ourselves like there might not be any tomorrow.

Nowadays we can get just about whatever we want whenever we want it. Not only that, we can choose to get our food from higher up the food chain and are eating loads more meat. Since it takes about 10kg of grain to produce 1kg of beef, this means that we're using even more land to feed ourselves and getting our food from further and further away.

But remember, if God didn't want us to eat fresh green beans in January why did he let us invent aeroplanes.

Bon appetit.

# Fifteen Supersize me

It takes 11,000 litres (11 tonnes) of water to grow the food a cow needs to make enough meat for a single quarter-pounder. In dry parts of the world this water will be pumped from underground, often using fossil fuels.

Cows are also well known for belching at one end and producing prodigious amounts of slurry at the other. The belches, from the first of their many stomachs, contain methane, a powerful greenhouse gas, and the slurry often runs off into streams and ponds giving a huge dose of food to algae which then quickly multiply and use up all the oxygen in the water. The end result is a load of ex-fish.

*What's the damage?*

Figures from Denmark say that producing 1kg of minced beef results in the release of 4.3kg of $CO_2$. So, you'd have to drive a Hummer for about 1km to match the 430g of $CO_2$ that comes from a single quarter-pounder (100g). If it cost £1, this would give a cost of £2.32/kg of $CO_2$.

*Can I be arsed?*

Use well known burger bars and the sums get a bit better. Then you can add in all the energy used to make the throw away plastic toys and to promote the brand, as well as that used to cook the burger and heat and light the restaurant. As a bonus, all that the packaging is more or less bound to end up as litter.

*Wind-up-ability*

Obviously, some brands have a higher profile with greenies than others. Maccy Ds aren't necessarily any better or worse than anyone else but, as a brand, they can be guaranteed to wind up your average yoghurt weaver something rotten.

## Sixteen Let your food do the walking

By and large, the further your food has to travel before it gets to your plate the more it stuffs up the environment. You can make the most of your "food miles" by choosing grub which hasn't only travelled a long way but has also done it by the most inefficient means.

*What's the damage?*

A plane carrying Mange Tout from Zimbabwe burns 0.83 litres of kerosene and emits 2.2kg of $CO_2$ for each 1kg pack. This is a total of 7.25kWh of energy. This is enough to boil a full 1.5 litre kettle over 50 times. In addition, the emissions are made straight into the upper atmosphere where they can have the biggest effect.

By contrast, stuff coming by boat uses pitifully little energy. For example, bringing a 5 kg Chicken Thailand by ship only uses 1.4kWh and only emits 0.4kg of $CO_2$.

*Can I be arsed?*

The alternative to eating what you want when you want it wherever it's come from, is to eat locally grown food when it happens to be in season. Rather than just going down the supermarket and doing your regular shop, you'd have to think what on earth your going to do with all the broad beans that turn up one week or Victoria Plums the next. It's no wonder so many of us take the easy option.

*Wind-up-ability?*

Unless you have the misfortune to live with a yoghurt weaver, nobody but the bin man is likely to notice.

**Seventeen** Veggies are for wimps, meat is for heroes.

If God didn't want us to eat animals why did he make them of meat?

The food chain starts with plants, which get their energy from the Sun, goes on up through the animals that eat the plants and ends up with animals that eat the animals that eat the plants. At each stage a whole load of energy goes missing. For example, it takes at least 7kg of grain to produce 1kg of cow and God knows how many kg of cow to make 1kg of lion.

To maximise your environmental impact it makes sense to aim as high up the food chain as you can. If you see any lion burgers in the shops then snap them up quickly, you may never get the chance again.

*What's the damage?*

While it takes 7kg of grain to produce 1kg of beef and 4kg of grain to produce 1kg of pork, it only takes 2kg to produce 1kg of chicken.

So, to maximise your stuff up potential go for red meat rather than white.

*Can I be arsed?*

Piece of piss

*Wind-up-ability*

Unless you live in a house full of veggies, hardly anyone will notice.

# Eighteen Grow your own

Back in the olden days dope used to come from all over the world. Nowadays it's as likely to come out of your neighbour's loft.

All you need to grow your own are seeds, compost, water and light. However, while there's loads of light outdoors, you wouldn't want your gardening to come to the attention of the plod. Hence, you'll need to set up shop indoors and use bloody great lamps instead.

Since the lamps use electricity, as well as getting stoned you'll also be emitting a whole load of $CO_2$.

*What's the damage?*

One Canadian web site says that a 1kW metal halide lamp is about right for two plants. The plants reach maturity in about 70 days and the lamp will be on for about 16 hours a day. Each plant can produce about 300g of cannabis (just over 10 ounces).

The total electricity you'd use would be about 1120kWh (1kW x 16hrs x 70days). Since each kWh of electricity produces about 0.43 kg of $CO_2$ you'll be emitting about 480 kg of $CO_2$. This works out at 24kg per ounce. (480kg/20oz)

At 7p per kWh this'll cost a total of £78.40, or £3.92 per ounce of dope.

*Can I be arsed?*

The biggest problem is having a secret place to do it. The trouble with lamps is that they produce far more heat than light. If you set the system up in a loft you run the risk of giving yourself away. Either to the thermal imaging cameras that some councils use to check up on how well insulated the houses are in their area, or if the snow on your roof melts long before it does on your neighbour's. In either case, the first visit you might expect is from the Drug Squad.

*Wind-up-ability*

Because it's illegal you can hardly go around bragging, except to your mates and, unless you rip them off, they're not very likely to get on your case, are they?

# Spend, spend, spend.

## Blunt Speaking

What's the point of having dosh if you aren't going to spend it? You may have noticed that after 9/11 the big worry in the States was that people would stop spending as much, factories would go out of business, credit card firms would lose their profits and the economy would nose dive.

Provided you avoid things where the biggest cost is labour, like original works of art, it's a pretty safe bet that any money you spend will result in extra waste in your bin, extra emissions from some factory and bigger holes in the ground (from where they dug out the raw materials).

To maximise your stuff up potential you might even consider spending money you haven't really got. The money will have been spent, the waste will be in the bin, the greenhouse gases will be working away in the atmosphere and even if you go bankrupt in a couple of years you can start all over again. However, this is best avoided if you own your own home because everyone needs a roof over their heads.

But remember, even if you're such a tight arse that you fill your house with energy saving light bulbs and other energy efficient gubbins, it'll just leave you with more money to spend somewhere else.

## Nineteen  Bling, bling, isn't gold brilliant.

Gold is a very special metal. It stays bright and shiny because it doesn't react with the air and therefore tarnish. This makes it perfect for the electrical contacts inside your i-pods, phones, laptops and PCs.

Once upon a time you'd look for gold by standing around in a river and swilling the muck from the bottom round and round in a pan until you were lucky enough to find a glittery bit.  Nowadays, you dig big holes, crush up loads and loads of ore and extract the gold using arsenic and mercury.

*What's the damage?*

Gold is valuable because it's rare and because it's rare you have to process lots and lots of rock just to get the teeniest bit. Extracting the gold for just one 18-karat ring produces about 20 tons of toxic waste and burns up a whole load of fossil fuels in the process.

*Can I be arsed?*

The only real problem with gold is affording it in the first place. Once you've got it it keeps its value pretty well and you can always pawn it if you find yourself short of a few readies.

Even if your crib is round at your mum's, get blinged up and you can still cut a dash in the 'hood.

*Wind-up-ability*

The whole point of gold is to show off and the whole point of showing off is to piss off those who can't. A genuine no-brainer.

## **Twenty** Become a diamond geezer.

Diamonds are pretty because of the way they refract light, and expensive because they're rare. Amazingly, they're made of the same stuff you'd find in your coal scuttle or in the lead of a pencil, carbon. But this is carbon that has been squeezed at enormous pressures and tremendous temperatures deep inside the Earth. By mimicking these conditions you can actually make artificial diamonds, but these are small and only much use for industrial applications like making abrasive surfaces for cutting tools.

Because they're rare, you have to dig a very deep hole just to get hold of not very many.

By becoming a diamond geezer you can bring a bit of sparkle to your life, let the world know you're loaded and give climate change a quick kick up the backside, all at the same time. Excellent.

*What's the damage?*

The world's largest diamond mine, the Orapa mine in Botswana, digs up about 16 million carats (3600kg) of diamonds every year. To do this it processes 20 million tonnes of ore and shifts a further 40 million tonnes of other rock just to get at the ore. So, each carat of diamond (0.2g) results in 3.75 tonnes of waste.

Not only does it take lots of energy to shift all this stuff around, it also takes loads to crush the ore and extract the diamonds.

*Can I be arsed?*

De Beers are the biggest diamond merchants in the World. They make sure that diamond prices stay high by controlling how many they let onto the open market. Unless they have a dramatic change of heart, the diamonds you buy will keep their value and, if things get a bit tight, you can always liquidate your assets (i.e. sell the blighters on e-bay).

*Wind-up-ability*

The conspicuous display of wealth always pisses some people off. Since these are precisely the people you'd want to piss off, getting blinged up is the perfect way to do it.

## Twenty one Thwack, thwack, thwack it's a jet-ski

Should you happen to live near water you can do your bit for climate change and annoy loads of people trying to enjoy a quiet day out, all at the same time. All you need is a jet-ski.

*What's the damage?*

A Yamaha jet-ski at full throttle burns 58.6 litres of petrol an hour. At 2.31kg per litre this means it throws out 135kg of $CO_2$ an hour.

At 90p a litre this'll set you back £52.74 and give a cost of 39p/kg of $CO_2$.

This may be twice the cost of leaving a light on, but it's at least twice as much fun.

Unfortunately, jet-skis don't come cheap. You'd be lucky to rent one for less than £25 an hour, so this would shove up the total cost to about £75, which works out at over 55p/kg of $CO_2$

*Can I be arsed?*

As well as the dosh you'd also need to wear a wet suit. Should you happen to be on the lardier side of manly, this might be a bit embarrassing.

*Wind-up-ability*

Jet skis not only make a lot of noise but they also make it in just the places where boring old farts go to for peace and quiet.

A soggy minded American researcher even went so far as to interview people on the shore and asked them how much they'd be prepared to pay for the jet-skis to go away. At a popular ocean beach this went as high as $538 per hour. He estimated the total noise cost to beachgoers in the USA as $908million per year. Go for it dudes.

## Twenty two Take to the skies

Fancy a weekend in Barcelona, a couple of nights in Bali, a stag night in Prague or a shopping trip to Fifth Avenue, no probs.

The great thing about cheap flights is, well, that they're cheap.

The other great thing is that they put their emissions straight where they have most effect, directly into the upper atmosphere.

*What's the damage?*

Each mile you travel by plane produces about 0.29kg of $CO_2$. Because this is put into the upper atmosphere, along with a load of heat reflecting water vapour, it's estimated to have at least three times the effect than if it were released on the ground. So, the effective stuff up potential is about 0.57 kg per mile.

To give an example, a flight from London to Rome is about 890 miles and would emit the equivalent of 507kg of $CO_2$. If the flight cost £30 this would give a cost of about 6p/kg of $CO_2$. (£30/507kg)

Blow me over with a feather if this isn't the best value yet.

*Can I be arsed?*

Depends whether you're scared of flying and how much you like airports. But, if you've got the time and fancy a treat then give it a go before the killjoys work out how to tax aviation fuel.

*Wind-up-ability*

Even if you can't be arsed to actually make the trips, you can wind up the yoghurt weavers by simply name dropping all the places that you've supposedly been to. "Isn't Las Ramblas wonderful", "Once you get used to the lard, the food in Hungary is brilliant", "There really is a lady boy on every corner"

## Twenty three Treat yourself to a big plasma telly

You've heard of solids, liquids and gases, the three states of matter. Well, there's a fourth, plasma. It's what most stars are made of and is what you get when you heat a gas up so much that its atoms lose control of their electrons and the little buggers escape. They don't do it for long, however, because the atoms snatch them back. But, each time an atom does this it chucks out a little bit of light. This is what makes a plasma telly work.

Of course, the amount of power a telly uses depends on how big it is. So, to compare different sorts (Rear Projection, old style Cathode Ray Tube, LCD and Plasma) you have to take the size into account as well.

*What's the damage?*

For each square inch of screen, a projection telly uses about 0.15W, an LCD telly about 0.25W, an old fashioned CRT about 0.30W and a Plasma telly about 0.35W. So, a 50 inch Plasma telly (diagonal 50 inches and area of about 1200 square inches) will use about 420W (and perhaps 20W on standby).

If you watched this telly for 4 hours a day, and left it on standby for the rest, your annual energy consumption would be about 210kWh. This would release about 91kg of $CO_2$ at a cost of about 23p per kg. Of course, if you haven't nicked the telly you'll have to include its cost in your sums as well.

*Can I be arsed?*

Watching telly is the definition of not being arsed to do anything else. So, of course you can.

*Wind-up-ability*

Young Theo Walcott recently fitted out his crib with 11 plasma tellies; one for every room in the house. Unless you can afford that sort of ostentation hardly anyone will notice the effort you're making. However, if it's a big enough screen there won't really be anywhere you could put it in your living room without half the neighbours being able to watch it for themselves anyway.

## Something of the Knight

That Dave, eh? Always in a hurry, always wants it now, whatever it is. He just doesn't do delayed gratification. He sort of knows it'll all end in disappointment, that the thing he thought would change his life will slowly turn into tomorrow's embarrassment. If you got him a new phone for Christmas he'd think it was out of date by Easter. He just can't bear the thought of things staying as they are. Without change what's he got to look forward to?

Now, while Dave likes change, Nature's a bit iffy about it. Some things, like bacteria, thrive on rapid change, others, quite frankly, are stuffed. The thing that makes the difference is how quickly they reproduce. Bacteria can go from birth to becoming grandparents during a single episode of Eastenders. An elephant, like a person, would have to wait at least thirty years.

This is important because the only chance evolution gets to interfere is when things reproduce. All that matters is that your sprogs live long enough to have sprogs of their own. Any mutation which happens to give more grand-sprogs is likely to spread (as the grand-sprogs get grand-sprogs of their own). Just like what's happened with anti-biotic resistant bacteria.

Even when evolution does interfere, it only does it a bit at a time and it does it at random. Of course, the more often you toss a coin the more likely you are to get heads. So, things that reproduce quickly get more tosses of the evolutionary coin and more chance of stumbling on a helpful mutation. One that helps them to get by in changed circumstances.

If you can't evolve fast enough you can always try moving to somewhere more comfortable. You can see how animals might do this. Even if they moved home by just a mile a year then, since the start of the industrial revolution (in about 1750), they could have moved from London to Newcastle.

The real trouble comes if change happens too quickly for you to be able to move, or if you run out of places to move to. Suppose, for example, you're an alpine type plant living high up on a tropical mountain. If the climate warms you just move up the mountain. But, eventually there isn't anywhere else to go and you're stuffed. Likewise with Polar Bears. Once their habitat of floating ice has melted that's it. No habitat, no bears. Except, of course, the terminally depressed ones stuck in zoos.

Then again, it's reckoned that 99% of the species that have ever lived are now extinct and that the average life time for a species is only about 2.5 million years. So, give it just a little bit of geological time and you'll barely notice that anything has happened.

## Twenty four Central heating on a stick

You've had your garden made over. You've decked the life out of any plants that might have had the nerve to try and grow in it. You've done your best to turn it into an outdoor room, but for most of the year it's bloody freezing. Time to get a patio heater.

Not only would you be emitting loads of $CO_2$, you'd also be dumping extra heat straight into the atmosphere. Double global warming or what?

*What's the damage?*

A typical patio heater is rated at 13kW, about the same as a regular gas boiler. It's like a central heating system on a stick.

Unfortunately, bottled gas doesn't come as cheap as the stuff straight from the pipe. An 11kg refill of propane will cost you about £20 and deliver about 150 kWh of heat (13.9kWh per kg). This gives a cost per kWh of just over 13p. Burning propane produces 0.23 kg of $CO_2$ for each kWh, which comes out at about 57p/kg of $CO_2$.

If the patio heater cost £100, lasted 5 years and you got through 2 cylinders of gas a year this would give a total cost of £300 (£100 + 2 x £20 x 5years) and emit 345kg of $CO_2$ (150 x 0.23 x 2 x 5). This works out at 87p per kg.

*Can I be arsed?*

All that sitting around outside. Could you really bear it? Especially when there's an unwatched plasma telly warbling away to itself in the living room. You could always get pissed. In which case you'd have to add the cost of the booze. Plus, those cylinders don't change themselves.

*Wind-up-ability*

Even if you never use the bloody thing, a patio heater to a yoghurt weaver is like a red rag to a bull. Maybe you should just get a dead one from a car boot and plant it out front as a wind up.

**Twenty five** Cover it in concrete.

The key component of concrete is cement and it's main ingredient is limestone (Calcium Carbonate). When you process this to make cement there's a chemical reaction that gives off loads of $CO_2$. As much again comes from the fossil fuels used in the processing.

Concrete over your front garden and you not only get a safe place for your motor, you also stop any pesky plants growing and do your bit towards local flooding by making sure that heavy rain runs straight off.

*What's the damage?*

Concrete costs about £100 per cubic metre. Covering a 20 square metre front garden to a depth of 20cm would need about 4 cubic metres (20 x 0.2) and cost about £400.

Concrete has a density of about 2400 kg per cubic metre, so this lot would weigh about 9.6 tonnes (1 tonne = 1000kg). Producing a tonne of concrete releases about a tonne of $CO_2$, so you'd emit about 9.6 tonnes of $CO_2$ at a cost of about 4p per kg. (i.e. well cheap, even cheaper than taking a cheap flight)

Sadly, as concrete ages it slowly reabsorbs the $CO_2$ that the limestone originally gave off. But, this takes a very long time and meanwhile it'll be out there in the atmosphere doing your dirty work for you.

*Can I be arsed?*

Once you've concreted over your garden you can forget about gardening. What's not to be arsed about that?

*Wind-up-ability*

A report to the London Assembly suggests that an area equivalent to 22 Hyde Parks has been paved over like this in London alone. Since this represents 2/3rds of all the front gardens in the capital then, even think about complaining, most of your neighbours won't have a patch of grass to stand on.

**Twenty six** My chopper's bigger than yours.

If you've made it, and really want to show you've made it, what could be better than getting yourself a helicopter. Just imagine it. Bumble out of your mansion, clamber into your chopper and whisk yourself away. Stuff the traffic jams, give two fingers to the speed cameras, just go straight where you want to go.

*What's the damage?*

As the old saying goes, if you need to ask you probably can't afford it.

A Bell Jetranger carries four passengers, cruises at 120mph and burns 114litres of fuel an hour. Helicopters have gas turbine engines burning kerosene. This produces 2.6 kg of $CO_2$ for each litre. So, a one hour flight would take you 120miles (193km) and emit 296kg of $CO_2$. This works out at 652g/km and is only about twice as much as a big SUV travelling at half the speed. There again, if a helicopter weighed as much as an SUV and had the aerodynamics of a brick you probably couldn't get the bugger to fly.

The big bonus for a helicopter, however, is the amount of noise it makes and how far this noise spreads. An early morning trip across the Smoke could easily disturb the sleep of a couple of million people.

*Can I be arsed?*

If you've got the dosh, this really isn't much of a question. If you haven't, then it really isn't much of one either.

*Wind-up-ability*

Having your own chopper can piss off loads of people and make them envious, all at the same time. What more could you ask?

## Twenty seven Give yourself an air-do with an air conditioner.

Not only is global warming bringing hotter summers, but your house gets even warmer because of all the new electronic gizmos you've got sitting there pumping out heat. So, say goodbye to sweaty arm-pits and get yourself an air conditioner.

*What's the damage?*

Air conditioners take air from a room, use the same principles as a fridge to cool it down, and then pump it back out again. Heat is a form of energy, and even the most efficient air conditioners only remove about three times as much energy as they use to do it.

So, if your plasma telly is chucking out heat at the rate of 300W the air conditioner will be using at least another 100W to get rid of it.

You can rent an air conditioner that uses 1.7kW of electrical power and produces 4.1kW worth of cooling for about £90 a week. If you had it on all week for 12 hours a day it would use 143kWh of electricity, produce 61kg of $CO_2$ (0.43 kg/kWh) and cost a total of £104.30 (£14.30 for the electric (143 x 10p) + £90 for the hire). This gives an overall cost of £1.71/kg of $CO_2$. Not cheap, but undoubtedly cool.

You can always make it work that bit harder by leaving the windows open so that there's a constant source of warm air for it to cool down. Alternatively, buy a crappy one which doesn't cost as much in the first place but uses more power to run.

*Can I be arsed?*

If, like Wayne Rooney, you need a constant electrical hum to get a good night's kip then this wouldn't be any sort of problem.

*Wind-up-ability*

In the good old USA there's only a short period in the spring, and another in the fall, when people feel able to do without either their central heating or their air conditioning. So, except in the poorest 'hoods, nearly every house hums nearly all year round. Hanging a bloody great air conditioner out of your front window, is a great way of showing that climate change isn't going to bring you out in a sweat.

# Twenty eight Take up power boating

Boats are a really efficient way of getting about, as long as you don't try to go too quickly. For example, the early canals gave the industrial revolution a right kick up the arse because you could shift 50 tons of coal using just a single horse.

But, as soon as you start going quickly the power you need starts to rocket. As a power boat enthusiast has said "It's like trying to push the boat uphill through wet cement".

*What's the damage?*

At full power a 320hp engine uses about 115 US gallons of fuel an hour. Since 1 US gallon = 3.79litres this is 437litres an hour. Burning 1litre of petrol emits 2.31kg of $CO_2$ so, at full power, you'd emit 1009kg of $CO_2$ an hour (i.e. just over 1tonne). If the boat's max speed was 63mph (100kph), you'd travel 100km and emit $CO_2$ at a rate of 10.9kg/km. (i.e. 10090g/km). This is about the same as 25 Range Rovers.

In another example, a modest sea fishing power boat (a Redfinn 10000) has two 370hp motors, a fuel capacity of 1,220litres and a stated range of 926km. This works out at 1.32litres/km and 3.04kg of $CO_2$ per km, about the same as 10 Range Rovers. Given that the cruising speed is only 46kph (less than 30mph) you can imagine how much it would use at 60mph.

Actually, since the resistance of the water goes with the square of the speed, you have to work 4 x harder to go twice as fast.

*Can I be arsed?*

Power boating is an expensive hobby. You can do your little bit by making sure you have a quick trip round the bay whenever you're at the seaside.

*Wind-up-ability*

As well as burning prodigious amounts of fuel, power boats (except those used by the military, who need to sneak up quietly) tend to be noisy. Since sound travels really easily over water, you're bound to piss someone off.

# Chuck it out?

## Blunt speaking

It's often said that we live in a throwaway age. Too right we do. What's the point of getting the latest stuff if your crib is still clogged up with yesterday's. Take tellys, for example, once you've got one in every room where the stuff are you supposed to put another one. The micro hi-fi was invented in Japan, where the houses are very small, so it could fit on the unused space on top of the fridge.

Nowadays it's reckoned that over 30% of the electrical stuff we throw out is in perfect working order and another 30% could easily be repaired. Sadly for the yoghurt weavers, it's usually a damn sight cheaper just to get another one. After all, you can't expect the local repairman to live on the sort of wages they pay in China.

When we buy something, we're only seeing the tip of the iceberg. For example, to make a telly you've got to extract all the different raw materials. Then you've got to use energy to process it and energy to shift it around. All this generates waste. It's been estimated that only 5% of the material that's been processed actually ends up in the final product. Because many things we buy are disposable, within 6 months of purchase only 1% of all the stuff consumed is still in use.

But, and there's always a but, some things use more when they're used than when they got made. So, while most of the time you can maximise your stuff-up-potential by chucking stuff out, sometimes it's best to carry on using it until the bitter end.

## Twenty nine Chuck out your old telly

Isn't telly brilliant, it's just a pity there isn't anything worth watching. Luckily you don't have to, just turn it on first thing in the morning, leave it burbling away in the corner like a demented old granny and Bob's your uncle.

You'll have noticed, over the years, how tellys are getting bigger and bigger. As well as letting you ignore them from a greater distance, they use more energy to run and it takes more stuff to make them in the first place.

*What's the damage?*

Typically, over 80% of the greenhouse gas emissions come from actually using a telly, with the rest coming from when it was made and when it gets left on standby.

It's reckoned that making a typical telly uses up over 600kg of raw materials and creates many times this amount of waste. For example, a telly contains lots of copper and for each kg of copper you generate about 90kg of waste.

In addition, extracting copper from its ore makes loads of silica dust which can cause asthma and pneumoconiosis, and produces clouds of sulphur dioxide which is responsible for acid rain.

Back in the olden days, when petrol contained a compound of lead to stop it going bang too soon, the kids who lived near main roads were even thicker than they are now. If you're careless when you recycle an old telly you can end up releasing lead vapour from the solder on the circuit boards. This means that a telly can keep you thick twice. Once when you watch it and again when you throw it away.

*Can I be arsed*

What do you call a room without a telly? That's right, the toilet.

*Wind-up-ability*

Planting a perfectly good looking telly next to the wheelie bin is a great way of showing the neighbours that you've got an even better one indoors. It may even inspire them to follow your example and get a new one for themselves. But, since we're a nation of telly addicts, it won't wind up anyone who isn't already a dedicated yoghurt weaver.

## Blunt speaking

Stuff it, I mean, I've got 50 years left tops. We've been promised flooded coastlines, monster hurricanes and melting permafrost. Who wouldn't want to be around to see that lot happen?

Some peeps talk as though the world shouldn't ever change, but they forget that it always has. Besides, without change, natural selection doesn't have anything to work on.

Where would your MRSA be if we hadn't chucked antibiotics at every little sniffle.

Would we be here today if, 3000 million years ago, little blue-green algae hadn't "polluted" the atmosphere by taking in $CO_2$ and spewing out oxygen?

Rather than trying to keep things the way they are, we should embrace change and let nature get on with sorting out the mess. After all, she always has, so far.

## Thirty Throw your tinnies in the bin.

Believe it or not, the first serious fridges were made to keep Australian beer cold. The beer can, tinnie, gets its name because they were lined with tin to stop the contents making them rust. Nowadays, most beer cans are made of aluminium rather than steel.

Aluminium ore, bauxite, is one of the Earth's most common minerals. You extract the aluminium by melting the bauxite and then passing a big electrical current through it. So, aluminium tends to get made where electricity is cheap.

*What's the damage?*

A typical aluminium can weighs 15g. It takes 14kWh of electricity to make 1kg of Aluminium, which works out at 0.21kWh per can. But, over 60% of the world's aluminium is made using hydroelectricity and so only 40% (0.08kWh) comes from fossil fuels. At about 0.5kg of $CO_2$ per kWh this gives 0.04kg (40g) of $CO_2$ per can. So, making the aluminium for each can produces more than twice the can's weight in $CO_2$. There will, of course, be extra emissions associated with turning a lump of aluminium into an actual can.

Recycling an old can into a new can uses only 5% of the energy it would have taken to make it from scratch. So, by recycling your cans you waste 95% of your stuff up potential.

Not only that, it also cuts down on the collateral damage caused by mining the ore and getting rid of the waste.

*Can I be arsed?*

Can I be arsed not to be arsed to recycle my old cans? What sort of question is that?

*Wind-up-ability*

Near where I live there's this sad old bugger who spends his time picking up cans that have been left lying around and takes them off to be recycled. Paradoxically, the best way of stopping this sort of counterproductive thing from happening is to stick your cans in a regular bin along with the chip wrappers, crisp packets and little bags of dog shit. No-one will want to recycle it then.

## Thirty one Don't throw out those old light bulbs (yet)

If you swap your old telly for a bigger one then you're likely to be chucking out even more $CO_2$ than you did before. Plus you'll get the benefit of all the waste and pollution from when it got made.

But, some things, like ordinary filament light bulbs, use much more energy when they're being used than when they're made. You might think a light bulb is there to shed light on things, but only 5% of the energy it uses actually comes out as light, the rest is just heat.

*What's the damage?*

Right at the start of this joyous tome we saw how much you could stuff things up just by leaving a 100W light bulb on all day. If you'd replaced it by a so called low energy bulb you'd only have used about 20% of the energy and wasted a glorious stuffupportunity.

An ordinary filament bulb lasts about 1000hours, so if you swapped it for a low energy bulb you'd miss out on the chance to use 80kWh of electricity (80W saving x 1000hours = 80000Wh = 80kWh) and to release 34.4kg of $CO_2$ (80kWh x 0.43kg per kWh)

*Can I be arsed?*

Since all you have to do is nothing, of course you can.

At work, if there's a box of old bulbs hanging around then make sure you use all of them before you even think about low energy alternatives. Your bosses won't have a clue what you're up to.

*Wind-up-ability*

Some tree huggers reckon that the only place you should see a filament bulb is in a light bulb museum. Use them all up first and don't give them that pleasure.

## Thirty two Keep that old fridge off the scrap-heap

A fridge takes heat from inside and dumps it outside.  To keep the inside at a steady temperature it has to pump heat out just as fast as it leaks back in again.  Modern fridges are better insulated and have better door seals so the pump doesn't have to work as hard and doesn't use as much electricity.

Of course, you can make things hard for the fridge by leaving the door open, keeping it in a warm room or by covering up the black coils at the back; which are where the heat is dumped.

If you've got an old fridge that chugs away in the corner then just let it be. Once you appreciate what it's doing to help warm the planet, that annoying noise can become music to your ears.

*What's the damage?*

Fridge motors may not be very powerful but they're bloody persistent. A typical older fridge is reckoned to use 500kWh of electricity a year, compared to 300kWh for a namby pamby new one.

That extra 200kWh will chuck out an extra 86kg of $CO_2$ (200 x 0.43). This is about the same as driving 300km (188miles) in a Range Rover.

*Can I be arsed?*

On the one hand, you might find it a bit tricky if the old fridge is so crap that you're having to put up with warm lager and runny butter. Despite its musicality, you might even find the constant chugging just the tinsiest bit annoying.

On the other hand, with a bit of cognitive therapy, you can get to see these things for what they really are, your own little contribution to global mayhem. It's a small price to pay.

*Wind-up-ability*
To be honest, this is a purely domestic matter. No one else will give a monkey's.

**Thirty three** Dump that bike, why ride when you can drive.

Getting a bike felt like a good idea at the time. Bollocks. Sweaty, red light jumping, self righteous, lycra clad toss pots, who'd be one of them? A right divvy that's who. Why waste time riding when you could be driving? The only pedal worth pushing is the one under your right foot, and it isn't one that goes round and stuffing round.

Even if you did want to ride the bloody thing, it's got a puncture, the gears are stuffed, the brakes are knackered (delete whichever is not applicable) and you can't be arsed either to mend the stuffer or to drag it off to a shop to get it mended. Just chuck it.

*What's the damage?*

Bikes are made from a mixture of different metals and alloys. The heaviest single bit is the frame and a typical aluminium one weighs about 3.5kg. Since it takes 14kWh of electricity to make 1kg of aluminium, and only 60% of this comes from renewable sources, this means that at least 8.4 kg of $CO_2$ were emitted just making the aluminium (3.5 x 40% x 14kWh x 0.43kg/kWh of $CO_2$). In addition to this, there'd be the energy used to shape it and stick it together.

As well as the frame there's the wheels, gears, brakes and tyres. Making all of these uses energy and produces waste.

*Can I be arsed?*

Crazy but true, you might be the sort of person who actually likes fixing things. You might think chucking something away well before its time is a wasted mending opportunity. Then again, probably not.

One little word of warning, though. Unless your careful about how you chuck it out, some saddo might rescue it before it gets to landfill. If you are going to dump it in a skip, make sure it goes in at the last minute. You can also have a bit of reckless fun by making sure the wheels are buckled and the frame is knackered.

Just think of the look of disappointment on the face of the yoghurt weaver when he finds that the nearly new looking bike in the skip is actually completely stuffed.

## Thirty four Buy a bike but never use it

Years ago I came across an article that claimed that bikes were worse for the environment than cars. Even I was a little sceptical at first, but when I read on I could see the point.

Most bikes get bought in a flush of misplaced enthusiasm. At the same time, the old wisdom that you get what you pay goes awol, and people buy cheap bikes which promptly start falling to bits. Far from being a bad thing, it's brilliant.

All those raw materials expensively dragged out of the earth, all that energy used processing them into something resembling a bicycle, completely stuffing wasted.

*What's the damage?*

Take the tyres, the 30% of rubber that comes from trees gives rise to habitat loss in tropical forests and produces pollution from pesticides and processing plants. The 70% from synthetic rubber uses up oil, consumes energy and generates waste. In addition, all this stuff gets hauled around using fossil fuels.

Then there's water. Making the tyres for a car uses 760 cubic metres of water, whereas making the metal uses only 450. Most of this water is used for cooling down the hot rubber when the tyre comes out of the mould. Obviously, these figures wouldn't be as high for a bike tyre but, even if you don't ride it, the tyres perish and have to be replaced anyway.

Finally, here's the killer maths. Take any number you like, except zero, divide it by zero and you get infinity. So, your big SUV might put out 350g of $CO_2$ per km, but an unused bike puts out an infinite amount per km.

*Can I be arsed*

Shopping's a way of life innit, and this is shopping with all the impression of good intentions (I'm going to get fit, I'm going to use the bike to get to work, I'll go out riding with the kids) but with exactly the opposite effect. Just remember that this is the age of irony.

*Wind-up-ability*

This is a tricky one. You could try telling a real cyclist about your wonderful new bike, its high tech features and all that, and watch what his face when it slowly dawns that you never actually ride the bloody thing.

**Thirty five** Buy a bike for your car.

Even better than buying a bike for yourself is getting one to stick on your car. Not only will you give the impression of being an outdoorsy sort of geezer, you'll also use up a lot more fuel.

At any decent sort of speed, the biggest force your motor has to deal with is air resistance. Ironically, you notice this even more on a bike when the only power you've got to deal with comes from your own legs.

Modern cars are designed to be aerodynamic. An easy way to stuff up the designer's good intentions is 1) to stick on a roof rack and 2) attach a bike to it.

*What's the damage?*

It's reckoned that a fully loaded roof rack can add 30% to your fuel consumption and a single bike might add 10-15%. But remember, the faster you go the bigger the payback.

Loading up a bike also gives even more excuse to make pointless trips to the countryside.

*Can I be arsed?*

Apart from attaching the rack and then lifting the bloody bike on and off, it's a real doddle. Luckily, some of these fancy new aluminium or carbon fibre bikes are surprisingly light. You wouldn't want to do it with your granddad's old three speed or the bargain bucket bike with full suspension that feels like its made from depleted uranium.

*Wind-up-ability*

This can be maximised by making sure you show due inconsideration to real cyclists before you storm past them in a cloud of exhaust fumes. Irony double plus.

# Thirty six Leave that loft alone

Because warm air tends to rise, the hottest air in a room is usually up near the ceiling. This means that houses lose lots of heat through the roof and a house that loses lots of heat is one that takes lots of heating.

Of course, houses lose heat through their walls, through their windows, through draughts and even through the floor, but the easiest, and cheapest, place to put in extra insulation is the loft.

Despite this, there are still plenty of houses in the UK that have little or no loft insulation. This means, if you're lucky, that you might be able to stuff things up by doing nothing at all. Just leave that loft, and all the junk you've put in it, to its own devices.

In Norway they call a poorly insulated roof "Central heating for Crows"

*What's the damage?*

This depends on how big your roof is, how warm you keep your house and whereabouts you live. A house on the chilly east coast of Scotland will lose more heat than one in balmy Cornwall.

The rate at which heat passes through each square metre of a material for each 1C temperature difference is called its U-value.

An un-insulated sloping roof has a U-value of about 2.01, but the same roof with 100mm of fibre glass insulation has one of only 0.33. If your roof had an area of 30 $m^2$, you set your thermostat to 22C and the outside temp was 2C this would mean that over a whole day the un-insulated roof would lose 28944Wh of heat (30$m^2$ x 2.01W/$m^2$/C x (22 – 2)C x 24hrs)). This compares to 4752Wh for the one with insulation. (i.e. 28.9kWh instead of 4.8kWh.)

If you were heating the house with gas, this extra 24.1kWh would release at least 4.6kg of $CO_2$ per day. Assuming the heating is used like this for an average of 3 months a year this would give an extra 418kg of $CO_2$ and cost about £88 (at 4p/kWh) Because gas boilers aren't 100% efficient the true figures would be even higher.

*Can I be arsed?*

Many people manage to do this without giving it a moment's thought. Even if I've spoiled it for you by pointing out how much it'll cost, you're still not very likely to get around to doing anything about it.

*Wind-up ability*

If it snows, then the snow will fall off your roof faster than those of your goody-goody neighbours and you run the risk of a visit from the drugs squad who'll think you might be growing your own in the loft.

**Thirty seven** Never wear more than just a T shirt.

What the flip's central heating for if it's not to keep you warm and what's a fancy T shirt for if no ones going to see it. Don't bury it under layers of clothing, just turn up the thermostat. Get some bright lights and you can pretend it's summer all year round.

*What's the damage?*

Turning up the thermostat by just 1C can easily add 10% to your bills and 10% to your $CO_2$ emissions. Bump it up by 5C and you could easily be paying, and emitting, half as much again.

So, if you currently spend £600 a year to keep the house at 20C and have to go around wearing jumpers you could crank the thermostat up to 25C and shell out an extra £300 a year for the privilege of going around in just a T shirt. You could probably save this by not buying any more outdoor clothing.

*Can I be arsed?*

The hidden bonus from just wearing T shirts is that you're much more likely to want to step straight out into the car than be arsed to walk anywhere. This is definitely win-win behaviour.

*Wind-up-ability*

The only way this'll wind up a yoghurt weaver is to invite the overdressed prat into your house and watch him sweat.

## Thirty eight Play golf in a warm climate

Go up to St Andrews, near Edinburgh, and you'll find a landscape of wind blasted heath and sandy hollows. The place looks like a bloody golf course without anybody doing anything to it all. Little surprise then, that this was where the game was invented.

But whilst your ordinary Jock might play golf, this isn't the same everywhere else in the world. In most places, it takes hard dosh to join a golf club and becoming a member is a real sign of having come up in the world. After all, it takes quite a few readies, and a lot of water, to turn a Spanish olive grove into an imitation of a Scottish heath.

*What's the damage?*

In Andalucia, that's the Costa del Sol and stuff, there are plans for over 200 golf courses. Every year, each one uses about 700,000 cubic metres of water. This is enough for the everyday use of about 15,000 people.

Take some intensively farmed land in the UK, turn it into a golf course and chances are that there'll actually be more bloody habitats for native critters than there were before. Do it in Spain and they're stuffed. Or would be if we could catch them....

But, the most important part of any golf course is the car park. This is where alpha males can display their affluence. There might be some sad so and sos who want to improve their handicaps, the rest just want to show off. The bigger, the shinier, the more expensive, the better.

*Can I be arsed?*

Of course, you do run the risk of actually enjoying the game. You might even get such a taste for the open air that you start taking walks in the country without having the excuse of trying to get a little ball down a hole. A good way of avoiding this particular temptation, is to keep your legs in box fresh condition by riding around on one of those little electric buggies.

*Wind-up-ability*

To be fair, it's not very high. Golf clubs are exclusive places and the only people that'll notice you're there aren't likely to give a monkey's.

## Thirty nine Leave hand tools to the handyman.

Be honest, the real point of DIY is trolling around B&Q fantasising about what you're going to buy next. After all, who wants an old fashioned saw that makes your elbow ache if you can spend a few quid on an electric one. A saw that not only does the job but also makes a satisfying buzz so that everyone can hear that you're working. It's also a lot more fun taking it out of the box when you get home.

*What's the damage?*

A typical little jig saw is rated at about 300W. Suppose it cost you £25, has a lifetime of 5 years and you use it once a month for about 20 minutes a time. You'll use 6kWh of electricity (300W x 1/3$^{rd}$ of an hour x 12months X 5yrs) at a cost of about 60p (at 10p per kWh) and release 2.58kg of $CO_2$.

If you include the initial cost this comes to £25.60 for a paltry 2.58kg of $CO_2$. At £9.92 per kg this is a bit steep.

But don't despair, there was also a load of crap given off when the thing got made in the first place.

*Can I be arsed?*

There's always stuff waiting to get done if only you could be arsed to get around to doing it. What better expression of intent than a brand new piece of kit. Even if it only makes it out of the box the once, at least you'll have made a start.

*Wind-up-ability*

Do your DIY in the middle of the night and you'll piss off the most mild mannered neighbour. But, merely possessing a grand collection of power tools isn't likely to raise very much attention.

## Forty Have a strimtastic afternoon in the garden

Back in the olden days, when the only sounds you heard in the garden were the clip-clip-clip of hedge trimmers, the swish-swish of a scythe or the gentle panting of the under gardener as he turned over the sods. Gardening in those days was a quiet activity. Nowadays you can make one of the most annoying noises known to man, the strimm, strimm, strimm of a strimmer.

As luck would have it, it hardly annoys the person doing it at all.

*What's the damage?*

A typical electric strimmer is rated at about 350W. Running this for an hour would use 0.35kWh of electricity and emit 0.15kg of $CO_2$.

An even noisier, petrol driven strimmer has a 28cc motor rated at 0.8kW, consumes 0.41kg of petrol per hour and emits 1.27kg of $CO_2$ (petrol has a density of 0.737 kg/litre and burning a litre of petrol emits 2.31kg of $CO_2$).

So, break free from the mains, get hold of a petrol strimmer and strim wherever you like. You'll be doing the dirties nearly ten times as quickly.

A Hectare (10,000 square metres) of pine forest can absorb about 1.5 tonnes of $CO_2$ per year. So, for each hour you play with a petrol strimmer you'd have to plant 10 square metres of pine forest to absorb the extra $CO_2$.

*Can I be arsed?*

If God hadn't wanted us to use strimmers, he wouldn't have let plants get so bloody untidy. Besides, if you gardened in the quiet old-fashioned way you might find yourself drifting into quiet contemplation of nature. That would never do.

*Wind-up-ability*

Truly excellent. Loads of annoying noise right where it's least wanted.

## Forty one Rake space for a leaf blower.

Autumn, a time of mellow fruitfulness, my arse. All those bloody trees scattering their knackered old photosynthetic units all over the place. Once upon a time you had to get out the rake and shift them by hand. Nowadays you just strap a leaf blower to your back and blow the buggers away.

*What's the damage?*

For a mere £254 you can buy a Husqvarna 145BT with a 40cc engine and an output of 2kW. It chucks out air at over 160mph, uses about 1 litre (0.74kg) of petrol an hour and in doing so produces 1.71kg of $CO_2$.

In a typical year you might use the blower for about 10 hours (half an hour a day for about three weeks) and emit 17kg of $CO_2$. To absorb this you'd have to plant about 110 square metres of woodland. About two squash courts worth.

But, even if you did plant the trees, it'd just give you more excuse to do it all over again next year.

*Can I be arsed?*

Leaves are a bloody nuisance. They sneak in the back door, block up your drains and, if you live where they've got the wrong sort, stop your train running on time.

Plus, if you leave them alone all they do is sit around making soil.

*Wind-up-ability*

This is where your leaf blower really scores. If you wanted to make a recipe for a noise nuisance then a small petrol engine would be one of the first ingredients. Just the job to take the "quiet" from the leafy suburbs.

**Forty two** Put down those clippers and get a hedge trimmer.

They may be good for your pecs, but old fashioned hedge clippers are just that, old fashioned. Not only that, since proper gardening is about showing nature who's boss it takes far too much effort to get the straight edged, artificial look that comes naturally with a hedge trimmer.

*What's the damage?*

A typical electric hedge trimmer is rated at 500W, uses 0.5kWh of electricity per hour and produces about 0.22kg of $CO_2$. A typical petrol one uses about 0.5litres of petrol an hour and emits about 1.1kg of $CO_2$.

If you used the hedge trimmer 5 times a year for two hours at a time then you'd emit either 2.2kg of $CO_2$ (electric) or 11kg (petrol).

But, while an electric hedge trimmer might set you back £30 a petrol one costs about £150. By now, I reckon you can do your own sums.

Assuming that your typical hedge is about 0.75m thick, that it absorbs about 0.15kg of $CO_2$ each year for every square metre and takes 10 years to reach maturity, then you'd need to plant about another 0.9m of hedgerow to absorb each kg of $CO_2$. (0.15kg per square metre = 0.1125kg for each metre of hedge = 1.125kg per metre over 10 years)

So to absorb the emissions from an electric trimmer you'd have to plant 2m of new hedge each year or, for the petrol one, a whopping 10m. It's not going to happen, is it?

*Can I be arsed?*

If you really insist on having hedges, and are too lazy to rip them out and replace them with something that never absorbed a bit of $CO_2$ in it's life, like a good brick wall, then you might as well just get on with it.

You could always add to your stuff up potential by loading up your SUV with all the trimmings and carting them off to the most distant dump.

*Wind-up-ability*

Bit of an arse this one. To most of your neighbours you'll just be doing your civic duty to tidy up nature. But in your heart, you'll know you're doing your best to stuff things up.

## Blunt Speaking

History. One damn thing after another, that's what they say. It reminds me of the Physicist who was asked what time was for and replied "it stops everything happening all at once".

Modern humans have been around for about 1 million years and, unless we do something really stupid, will be around for another million years to come. So, we're just about in the middle of history.

On that sort of timescale the odd thousand years is neither here nor there and a century is just a few ticks of the historical clock (less than 0.01%).

Now, you might claim that the most interesting bit of history was the rise of the earliest civilisations about 10,000 years ago but you could also argue that the most interesting bit of all is just about to start now.

Until the beginning of the Industrial Revolution we were stuck with renewable resources. We burned wood, peat or dung to stay warm, in Holland they had wind driven pumps, elsewhere there was the odd waterwheel, ships got blown along by the wind, ploughs, carriages and carts were pulled by bio-fuel powered animals.

Since then, we've been burning fossil fuels about 1 million times as quickly as they were originally laid down. Something has to give and, given all the history there's been and all that there's going to be, I want to be around when it does. In the long run it really won't matter exactly when the crunch comes, so it might as well happen now.

# Use the system

## Forty three Get into debt.

Money is wonderful stuff. If a nation starts running out, it just prints some more. Only trouble is, if you've got more money chasing the same amount of stuff then prices go up and you get inflation. But, if the overall size of the economy grows then you can avoid inflation by having more stuff for the extra money to chase.  Indeed, if you can make more stuff more cheaply then prices can even start to fall, just as they have for computers and other electronic gizmos.

When you get into debt, you're effectively paying for stuff today out of what you anticipate you'll earn tomorrow.  The debt can be paid off in two ways. Either you accept that you'll make do with less stuff in the future or you get yourself a wage increase so you can pay off the debt and still keep on consuming at the same rate.

Inflation really is a bit of an arse and most countries will do what they can to avoid it. So, one of the best ways of making sure that we keep on going for economic growth, keep on making more and more stuff, keep on emitting even more greenhouse gases, is to get yourself, and your country, into debt.

*What's the damage?*

Put six economists in the room and they'll give a dozen different answers, because they can't even agree even with themselves. Safe to say, they'll all agree that if the economy isn't growing then it's actually going backwards.

Even if you don't pay your bills, and go bankrupt, the economy as a whole still has to. If a bank writes off your debt you can rest assured they'll screw the money out of someone else.

*Can I be arsed?*

Again, it's more a case of can I be arsed not to? Just think of all the junk mail inviting you to get another credit card, of all the store cards stuffed in your wallet and think of Carol Vorderman's warm words on daytime telly.

To be honest, absolute pants. But you'll know you're doing your bit to keep the runaway train on schedule.

## **Forty four** Invest unethically?

Of course, what is or isn't ethical is a matter of opinion. But, for arguments sake, if we take ethical investment to mean avoiding arms companies, tobacco firms, big oil companies and the like, then unethical investment simply means the opposite.

So, if you've still got any readies left over put them into unethical investments and let them do your dirty work for you there.

*What's the damage?*

For example, take the arms trade. If you supply people with weapons then chances are they're going to use them. Even if they don't, if the other side gets newer weapons they'll just have to buy more to keep up. You may be able to do without the latest tractor, but you can't get away with last year's fighter if your enemy's got this year's surface to air missile. After all, it's called the "arms race" innit?

*Can I be arsed?*

The great thing about these investments is that they don't half pay well. Take tobacco, here we've got one of the most addictive substance known to man and you're allowed to sell it without any real questions being asked. All it takes is a bit of gentle promotion to get youngsters onto the low nicotine starter brands and then you've got customers for life (i.e. until death). The big companies don't bother to advertise the stronger brands because the customers desperately seek them out all by themselves.

*Wind-up-ability*

Get an "I support the Arms Trade" sticker for your SUV, an Exxon anorak, a Philip Morris International tie and a British American Tobacco baseball cap and you'll be winding people up all the way to the bank.

## **Forty five** Become a slum landlord.

Way back at the start of this book we looked at ways you could stuff things up around the home. The great thing about being a slum landlord is that you can stuff things up at someone else's expense.

For example, if the house you rent out doesn't have any loft insulation, if the water cylinder isn't lagged, if it has draughty doors and windows and if it has an inefficient old boiler, then your tenants will be burning loads more fuel than they have to, emitting far more greenhouse gases than they might and they'll be doing all of this out of their own pockets. Indeed, you'll be charging them for the privilege.

*What's the damage?*

Britain's houses are by far the least efficient in northern Europe. Go to chilly Scandinavia and you'll find that the houses there are so bloody well insulated that they get by on the heat from a couple of light bulbs and a warm dog sat in the corner. Meanwhile, our houses produce about a third of Britain's $CO_2$ emissions.

Increasingly tight building regulations are making it harder and harder to maintain the stuff up potential of new houses. But, the reason house prices are so bloody high is because we aren't building enough of them anyway. So, you can do your bit for global warming by helping to keep as many old houses as possible as crap as possible.

*Can I be arsed?*

There's stuff all extra rent to be got off an energy efficient house, but a load of extra expense in putting things right. You don't have to pay the bills so what's there not to be arsed about.

*Wind-up-ability*

Many years ago a slum landlord in London gave his name to naked profiteering from ill maintained properties. You might annoy your tenants, but you'll never be another Rackman. Nobody, apart from your skint tenants, is likely to pay much attention to this particular contribution to climate change.

# Forty six Drain a bog.

Wimps who don't want to stand up and be counted for their contribution to Climate Change often try to salve their petty consciences by "offsetting" their emissions. This usually involves giving money to someone who promises to plant a few trees to soak up your $CO_2$. Luckily for us, this doesn't always work the way the wimps would like it to.

For one thing, it takes time for trees to grow and meanwhile your emissions are out their doing the dirty. For another, trees can burn down and leave you right back where you started. In addition, these poor souls might even be making things worse (or better, as it were).

One, it turns out that trees absorb more heat from the sun than a field full of grass. So, unless you're planting trees near the equator, you can end up making the planet warmer than if the $CO_2$ they've absorbed had just been left to float around in the atmosphere.

Two, there's usually far more $CO_2$ locked up in the soil than there is in the vegetation that grows in it. This is particularly true of wetlands where a good peat bog can store several tonnes of $CO_2$ per square metre whereas the trees on top can only store a few kg. So, draining a bog to plant trees actually releases far more $CO_2$ than the trees could ever absorb.

*What's the damage?*

A report produced by the University of Edinburgh in 1999 suggests that draining deep peat releases 200g/m$^2$/year and draining lowland wetland releases 297g/m$^2$ /year. So, if you drained a hectare (10000 square metres) of wetland you could release nearly 3 tonnes of $CO_2$ every year. (10000 x 0.297kg = 2970kg = 2.97tonnes) This is about the same as running a big SUV for about 8000km (5000miles) or a return flight from London to Hong Kong.

*Can I be arsed?*

If you were doing it just for its own sake, then I doubt you could. But, if draining the land made it more valuable, to developers for example, then it'd be a no-brainer.

At the moment, very few people would spot what was going on and even fewer would connect it with Climate Change. Just wait a bit and, in a few years time when even the doziest yoghurt weavers catch on, you'll have them tearing their hair out in frustration.

## Forty seven Fight for your right to parking.

Intimidated by self righteous do-gooders and yoghurt weavers, loads of businesses have been obliged to produce so-called Green Travel Plans. To you and me, that means taking away the parking places of everyone except Senior Stuffing Management.

It goes like this. If people don't have somewhere to park they'll find "more sustainable" ways of getting to work. Stuff that for a lark. You're not getting me on a bus with a load of losers, pensioners and screaming kids, and I can't exactly see myself dressing up in lycra and puffing in on a bike.

*What's the damage?*

Remember, you're not just fighting for yourself. If you keep the car park it won't only be you that's doing your bit. It's not only you that'll keep on polluting and congesting but your work mates as well. A little bit of effort could end up going a long way.

*Can I be arsed?*

The trouble with this sort of thing is that it means sticking your head above the parapet and you run the risk getting it shot off. Best thing to do is to circulate a petition, making sure that your name isn't the one at the top.

In addition, imagine what an arse it would be if you really did have to find some other way to get to work.

*Wind-up-ability*

A perfect opportunity to wind up the odd sad eco-bugger that it's your misfortune to share an office with.

**Forty eight** Vote for the cynical greedy so and so party.

People, what are they like? Do they love their neighbours like themselves? Do they turn the other cheek? Is there a Mother bleeding Theresa inside each one of us itching to get out? My arse. Anyone who thinks we're in it for anyone but ourselves needs their little head examining.

Sadly, too many of us live under the delusion that if only we were nice to each other we'd all get on famously and the world would be a better place. They haven't quite twigged that unless you let loose your little inner greedy so and so then some other greedy so and so will simply step in to take advantage.

Successful political parties have always known this. Just before the elections in 1992, John Major was way behind in the polls but he still won. Turns out that even though it wasn't very popular to say you were going to vote Tory, once in the privacy of the polling booth the inner greedy so and so couldn't help but come out.

*What's the damage?*

Huge. Since the 1992 Rio Earth Summit our political leaders have all known that we're messing up the planet. But, they didn't trust us to vote for them if said it or if they did anything that might hit anyone in the pocket. Look how the Government caved in to the glorious fuel tax protestors in 2001. It took years before they dared to step back on the fuel duty escalator.

As long as they're more interested in being in power than in doing the "right" thing, this is how it's going to stay. Keep on being selfish, cynical and greedy, and the Cynical Greedy So and so Party is the one that's going to get elected. Whatever name it happens to go by at the time.

By being selfish together we can achieve far more than if we acted on our own. Is there a paradox in there somewhere?

*Can I be arsed?*

You will have to start paying attention, or at least get someone to remind you when there's an election. Then you'll actually have to get off your arse and vote.

If all the cynical selfish so and sos either stay at home, or forget to send in the forms, you run the risk that some genuinely well meaning tosser will get elected instead. Never mind, before too long exposure to all those politicians is bound to make them selfish, greedy and cynical just like the rest.

*Wind-up-ability*

While no-one can know who you've really voted for, you can announce your intentions with an inflammatory poster or two.

# Seeming to do the right thing

## Forty nine Run your motor on bio-fuel

Bio-fuels are the perfect "get out of jail free" card for people who love their motors but have begun to feel guilty about using them. What a. gloriously simple idea. Instead of using petrol extracted from oil, you use ethanol brewed from crops. Burning ethanol releases $CO_2$, just like burning petrol but, unlike petrol, you can claim it gets sucked up again by next year's crop. Conscience clear, planet saved, motor still on the road. Poor deluded divvies, when will they learn. It takes a bloody site more to get away with it than that.

For one thing, what sort of fuel do you think is used by all those tractors, fertiliser plants and bio-reactors? For another, there's only so much bleeding land to grow stuff on, and the Sun supplies it with only so much energy. Whether you like it or not, you can't get more out than gets put in. The end result is that you may well be emitting more $CO_2$ than if you'd just stuck to good old fashioned petrol.

*What's the damage*

The USA has a target to produce 10% of its vehicle fuel like this. It turns out this would use 30% of its agricultural land and that it takes as much land to fill the tank of an SUV once as it does to feed someone for an entire year. If there's a competition between the food for the world's poor and fuel for our motors you can guess who'll win.

Plus, as the demand for bio-fuels grows, the price will rise and it'll be worthwhile using more marginal land, using more fertilisers, clearing more forests, and draining more bogs in order to grow it. All of these things are likely to put more $CO_2$ into the atmosphere than the bio-fuel could ever save.

*Can I be arsed?*

If you live in the EU or the USA you won't have much choice. Both of them have set targets to produce more bio-ethanol. Even now, if you've got a diesel engine it's not a lot of trouble to pour in the odd bottle of cheap cooking oil.

*Wind-up-ability*

If you want to piss off real greenies then a "This car runs on bio-diesel" stuck in the back of a Range Rover will do it perfectly.  At best they'll think of you as a well meaning idiot and at worst as a self satisfied wind up merchant; which wouldn't be that far from the truth, would it?

## Fifty Get a wind turbine but forget to insulate the loft.

Want to make a highly visible but ultimately pointless gesture?  Want to demonstrate your "green" credentials to the man in the street? Then get a wind turbine for your roof.

Don't get me wrong. Those big turbines up on a hillside really do do the business. Even if you include the $CO_2$ that comes from making them and their concrete bases, they still only emits about 1/10 of what you'd get from a modern gas fired power station. It's the little ones stuck peoples' houses in cities that aren't up to much. Unless that is, you've got a pointless point to make to people who don't know any better.

For a start, to work properly they need a nice smooth flow of air, not the turbulent crap you get near buildings. The turbine simply can't respond to the constant little changes in wind direction. Secondly, the power they produce depends on the cube of the wind speed. This means that double the wind speed gives 8 times the power. Or, the other way round, halving the wind speed gives only 1/8 the power. Close to the ground, wind speeds are usually just too low.

*What's the damage?*

£1500 or so will get you a little turbine that's rated at 1200W in a wind speed of 12m/s. This is a pretty stiff gale. At a more typical 6m/s you'd only be getting 1/8 of this (150W); which is just about enough to run your fridge. If it was this windy for half the time you'd probably only generate about 650kWh of electricity a year (0.15kW x 12hours x 365days).

As you can see from **36** "Leave that loft alone", simply failing to insulate your loft could easily waste 24.1kWh a day. If you only had the heating on for three months a year, it would still come out at 2,200kWh and no one else would be any the wiser.

*Can I be arsed?*

This really depends on how much dosh you've got to waste. If the turbine lasts 15 years, didn't cost you anything to get it installed, never needed fixing and generated a total of 9,750kWh of electricity, it comes out at about 16p per kWh. This is more than if you'd just bought the stuff straight from the grid. If you'd put £1500 in the bank at 4% interest then after 15 years it'd be worth about £2700. You'd still have your original £1500 plus £1200 left over to stuff things up even more effectively.

*Wind-up-ability*

Combined with a big SUV parked in the drive, it's the perfect way to drive your genuine yoghurt weaver up the proverbial wall. Not only that, but you'll be in the good company of all those out there who want to make a point without really doing anything at all. This will probably include the leaders of whichever Cynical Greedy So and so Party happens to be after the "green" vote at that particular time.

## Blunt speaking     "Welcome to the Anthropocene"

Since the start of the industrial revolution we've seen off so many other species, and pissed about with so much of the Earth's surface, that we've been given our own geological era; the Anthropocene.

Maybe, one day, the next intelligent life form that evolves will watch "Anthropocene Park". The movie that shows what might happen if they brought us, homo sapiens, back to life. If you think dinosaurs are frightening, they've got nothing on us. They hung around for 60 million years and it took a major asteroid impact to see them off. We'll be lucky to last a million and we'll have no one to blame but ourselves.

When a geologist looks at rocks they can tell what era they came from by looking at the fossils they contain. After each of the Earth's previous mass extinctions, the surviving species rapidly evolved to fill up any newly vacated ecological niches. After a few million years everything tends to settle down again and its as though nothing had ever really happened. All that's different is that the rocks contain a different set of fossils.

We may not be around in a hundred million years time but we'll have left our tag on the geological record just as surely as any graff from the 'hood. Everything else may have gone, nature's great at claiming things back, but they'll still know we were here. Once.

So, what's likely to happen in the next few hundred years?

We might have a sudden outbreak of good sense and voluntarily choose to consume less and share more. We might decide that whatever the answer is it isn't "shopping" and we might start giving people status for what they do rather than what they possess. Then again, we might not.

Rich people will probably get even richer and find ways to insulate themselves from the worst aspects of any environmental crisis. But, they'll still need people to do the stuff for them that they're incapable of doing for themselves so they won't be able to cut themselves off completely.

No doubt the rich will still get the pleasure and the poor will still get the blame.

Good luck. Dave.

# Appendices
(things Phill thinks it's useful to know)

**Work, Energy and Power**

Part 1) Forces
Part 2) Basic definitions
Part 3) Basic units
Part 4) Derived units
Part 5) Electricity
Part 6) Electrical units
Part 7) Heat
Part 8) Heat definitions
Part 9) Heat transfer
Part 10) Heat calculations

**Sources of Energy**

1) Energy from the Sun
2) Energy from the Earth

**Energy in humans** (Why big brains are expensive)

**+ The wit and wisdom of Al G Bra**

# Work, Energy and Power

## Part 1: Forces

Pushes and pulls, that's all forces are. But it took a special genius to understand them well enough to be able to work out precisely what they do. The great Isaac Newton, of apple tree fame, saw right through the confusion of every day life and summed up everything about forces in just three simple laws. **Newton's Laws of Motion.**

The first two tell us what forces are and how big they are. The third explains the conservation of energy. The law which says you can change energy from one form to another but you can't actually create it or destroy it.

**Law one:** An object will keep on moving in a straight line at a steady speed unless an external force acts on it.

> So, if something speeds up, slows down or changes direction there must have been a force. In a football match, this means that if the ball is deflected then someone must have touched it.

Obviously, the bigger the force the bigger the effect. To deal with this Newton introduced the concept of momentum (how much motion something has). He said this just depends on its mass and how fast it's moving. The momentum is what you get when you multiply the mass by its velocity.

**Law two:** If you use twice the force the momentum will change twice as fast.

> At everyday speeds, where masses don't change very much, this becomes the familiar

### Force = mass x acceleration

**Law three:** If body A exerts a force on body B then body B exerts an equal but opposite force back on body A

> If I push against the wall then the wall pushes back just as hard on me.

We'll see what this means when we get to work and energy.

## Gravity:

Newton is also famous for his law of gravity. He said that every mass in the Universe attracts every other mass in the Universe and that the size of this force just depends on how big they are and how far apart they are. Make either mass twice as big and you'll get twice the force. Put them twice as far apart and you'll get ¼ of the force. 3 times as far apart and you'll get $1/9^{th}$ of the force etc.

Using these simple rules he was able to explain, in detail, how the planets orbit the Sun and how the Moon orbits the Earth.

Everybody on Earth gets pulled towards the Earth. We call this force our weight. The simple, but expensive, way to lose weight is to move further from the Earth's centre. Go up about 6,400km (and end up about twice as far from the Earth's centre as you are now) and you'll weigh about a quarter of what you do on the ground.

**Weight:** The weight of an object is the force exerted on it by gravity. If you take the object to the moon, where gravity is weaker, it will weigh less. Take it to Jupiter, where gravity is stronger, and it will weigh more.

Using Newton's second law you can see that the weight of an object is given by          Weight = Mass x local acceleration due to gravity

At the Earth's surface, the local acceleration due to gravity (g) is about 10 metres per second per second (m/s/s). If you drop a stone, then after one second it'll be going at 10 m/s, after two seconds it'll be going at 20 m/s etc. etc.

But, on the Moon, where gravity has about $1/6^{th}$ of the strength, the local acceleration due to gravity is only about 1.7 m/s/s.

### Part 2: Basic definitions

**Work** is something done by forces, but only when they move. The force that stops your elbows going through the table isn't doing any work because it isn't moving, but the force you use to push a supermarket trolley is.

The amount of work a force does depends on two things. How big it is and how far it moves. For example, pushing a supermarket trolley down two aisles takes twice as much work as pushing it down one. Likewise, pushing two trolleys at once, and using twice the force, takes twice as much work as just pushing one.

So, **Work = Force x Distance**

If something has **Energy** it means you could get it to do some work. The amount of energy it's got is just the amount of work it could do.

So, **Energy = capacity to do work**

Energy can get converted from one form to another. E.g. a light bulb converts electrical energy into heat and light, your body converts food energy into heat and motion. The more **powerful** the device the more quickly this happens.

So, **Power = rate of doing work = rate of converting energy from one form to another.**

### Part 3: Basic Units

The basic units of mechanics are those of mass, length and time. All the others, (velocity, acceleration, force, work, energy and power) are just different combinations of these three.

Once upon a time we had a whole load of different basic units. For example, the furlong (1/8$^{th}$ of a mile or 220 yds) was the length of a ploughman's furrow (furrow's length), a chain was 1/10$^{th}$ of a furlong (22 yds, the length of a cricket pitch) and an acre was an area 1 furlong long and 1 chain wide. The yard was the maximum distance between the King's nose and the tip of his finger. I still ask for a pound of cheese and know my weight in stones. Time used to be much more informal. For example, in some places, a small bit of time was called a "pissing while". Nowadays, scientific units are standardised across the globe. (except for engineers in the USA who still fart about with feet and pounds)

**Mass:** The basic unit of mass is the kilogram (kg) and there's a standard 1 kg lump of platinum-iridium in a laboratory near Paris. It was originally defined so that 1000 cubic centimetres of pure water weighed exactly 1 kg.

**Length:** The basic unit of length is the metre (m). Strange as it may seem, this is now defined by how far light in a vacuum travels in some tiny fraction of a second. If someone makes a more accurate measurement of the speed of light, then the length of the metre will change but not the speed of light. It will always keep the same value but the metre will shuffle to adjust.

**Time:** The basic unit of time is the second. Once defined by the length of the day (24 hours each divided into 60 minutes, each divided into 60 seconds) it's now defined by counting the natural vibrations of an atom of Cesium 133. At over 9 billion per second this is not as simple as it might seem.

Apart from electrical units, all the others are derived from these three.

### Part 4: Derived units

**Speed and velocity:** We're all familiar with miles per hour (mph), but the standard scientific unit of speed is the metre per second (m/s). Speed and velocity have the same units. The only difference is that velocity includes the idea of direction as well. So, to talk about velocity of an object you also have to talk about where its going. E.g. a plane is travelling at 120m/s in an easterly direction or I pushed you at 3m/s **off** the edge of the cliff.

**Acceleration:** If a car accelerates from 0 to 60mph in 10 seconds, then the acceleration is 6mph per second. Sticking to the basic units of metres and seconds gives the slightly more confusing unit of metres per second per second (m/s/s).

**Force:** From Newton's second law Force = mass x acceleration. This gives the units of force as kilogram metres per second per second (kg.m/s/s)

The force that's needed to give a mass of 1kg an acceleration of 1m/s/s is called the **newton** (N). Using Newton's $2^{nd}$ law this means a force of 10N would give a mass of 1kg an acceleration of 10m/s/s and a mass of 5kg would have a weight (on Earth) of about 50N
(weight = mass x g  = 5kg x 10 m/s/s = 50 N)

**Work:** The basic unit of work is the joule (J). This is the work done by a force of 1 newton when it moves through 1 metre. So, a force of 10N moving through 50m will do 500J of work.

**Energy:** Has the same units as work, joules. A 50N weight falling through a height of 10m could do 500J of work. (Work done = Force/weight x distance/height). The fact it could do this much work meant that it had 500J of potential energy.

**Power:** The basic unit of power is the watt (W). If something has a power of 1W it means it can do 1 J of work every second. 1kW = 1000W. A 100W bulb uses 100J of electrical energy every second.

**KWh:** Another unit of energy is the kilowatt hour (kWh). This is the amount of energy used by a device running at 1kW (1000W) for one hour. (3600s). So 1kWh = 1000W x 3600s = 3,600,000J

### Part 5: Electricity

Electricity is funny stuff and it took a long time to figure out what it was. You may not be able to see what it is, but you can certainly see what it does. After all, you'll have felt it if you've stuck a 9V battery on your tongue or played with a bit of grass and an electric fence, and if you've been out in a thunder-storm you'll have seen the flash and heard the thunder. You'll have noticed what electricity can do but you won't be any the wiser to what it actually is.

At first, scientists studied static electricity. This is what you get when you rub a balloon on your jumper and is what causes all that dust to gather on your telly. The explanation they came up with involved the idea of electrical charge, which could be either positive or negative, along with the simple rule that like charges repel each other but unlike charges attract.

**Atoms:** It turns out that ordinary, everyday, matter is made of atoms. Each atom has a positively charged nucleus surrounded by a cloud of negatively charged electrons. The negative electrons don't wander off because they're attracted by the positive nucleus. The atom as a whole is neutral because the positive and negative charges cancel each other out.

**Static electricity:** Some materials attract electrons more strongly than others. Rub two different materials together and the one that likes electrons most will snatch a few from the one that likes them least. The one that likes electrons most will end up with a slight negative charge and the one that likes them least will be left with a slight positive charge.

**Current electricity:** Some materials let electrical charge flow through them. If you make one end positive and the other negative then charges will flow until the balance is restored.  In metals, it turns out that the atoms share some of their outer electrons and it's these that can move and carry the current.

### Part 6: Electrical units

Electrical charge is measured in **coulombs** and the unit of electrical current (the **ampere**) used to be defined as a **coulomb** per **second.**

Nowadays, we can measure currents much more precisely than charges. So a **coulomb** is now defined as the amount of charge which flows by when a current of **1 ampere** flows for **1 second.**

Voltage is the electrical equivalent to pressure. The higher the voltage across a conductor the more current will flow through it.  Similarly, the higher the water pressure in a shower the faster the water will come out.

Just as high pressure water is more energetic (you could get it to do more work) so is high voltage electricity. In fact, **Voltage** is defined as the amount of energy carried by each coulomb of charge and 1volt = 1 joule per coulomb.

For example, a 12V car battery gives 12J of energy to each coulomb of charge that it sends out into the car's circuits.

**Electrical Power.** Remember, power is a measure of how much energy is converted every second. So, electrical power = energy per coulomb x coulombs per second. This is the same as **voltage x current**.

E.g. If a 12V car headlight rated at 60W it will draw a current of 5A, because 60W = 12V x 5A

# Part7: Heat

We all know that heat flows from warmer things to colder things, but do we know what it actually is?

Well, another name for it is "internal energy". It's the energy that something has not because you've put it on a high shelf (gravitational potential energy), not because you've stretched it like a spring (elastic potential energy), nor because you've thrown it at someone (kinetic energy), but because the atoms it's made are more energetic.

How can this be? Well, unless they are at absolute zero ( -273C ), atoms are in constant restless motion. In a gas they fly about all over the place banging into each other and the walls of their container. In a solid they can't dash about but they can vibrate and, as the solid expands, push each other further apart. In a liquid they don't only move about, they also push each other further apart and even spin on the spot. The higher the temperature the livelier the motion.

For example, if you wanted to increase the kinetic energy of the air atoms in a football you could either kick the ball, in which case the atoms would all move off in roughly the same direction, or you could put the ball in the oven and warm it up, in which case the atoms would move about more quickly but in completely random directions.

It turns out that if two objects have the same temperature then their atoms have the same average amount of this random kinetic energy. When a warm object touches a cold object the atoms on their surfaces come into contact. Because the warm atoms have more energy they hit harder than the cold atoms, do more work in the collisions and pass over some of their energy. For example, a cup of tea may be hotter than a bath full of water but it doesn't contain anything like as much heat. Pour the tea into the bath and the bath's temperature will barely change. The amount of heat energy something has not only depends on its temperature but also on how big it is and what it's made of.

**Absolute temperature:** While we usually measure temperature in degrees celcius or fahrenheit, scientists prefer to use the absolute temperature scale (Kelvin). This still has 100 degrees between the freezing and boiling points of water, like the celcius scale, but starts from absolute zero. So, in the Absolute Scale, water freezes at 273.16K and boils at 373.16K.

## Part8: Heat definitions

**Heat capacity:** The amount of heat it takes to change the temperature of an object by 1K is called its heat capacity and is measured in joules per kelvin. For example a 2kg brick has a heat capacity of about 1680 J/K. whereas a 2kg lump of gold would have one of only 258J/K.

**Specific heat capacity:** This is a measure of how much heat it takes to change the temperature of 1kg of a particular material by 1K. For example, the specific heat capacity of copper is 385J/kg/K whereas that for water is a walloping 4200J/kg/K. Indeed, water needs more heat to change its temperature than anything except liquid ammonia. This is why it's really useful for shifting heat from one place to another.

**Latent heat:** It takes energy to change the state of a material from a solid to a liquid or from a liquid to a gas. For example, the water molecules in liquid water are much closer than they are in steam. So, to turn water into steam it takes energy to do the work to pull the molecules apart. This extra energy is called the latent heat and is called latent because you get it back again if the steam condenses; as those of you who've scalded themselves with steam from a kettle can testify.

If you're going from a solid to a liquid it's called the latent heat of fusion and from a liquid to a gas it's called the latent heat of vaporisation. It's measured in joules per kilogram.

For example, it takes 540,000J of heat to melt 1kg of ice at 0C to form 1kg of water at the same temperature. I.e. the latent heat of fusion is 540,000J/kg. Similarly, the latent heat of vaporisation of water is a whopping 2,300,000J/kg. This means it takes more than five times as much heat to turn 1kg of water at 100C to steam as it took to raise its temp from freezing point to boiling point ( 2,300,000J v 420,000J)

## Part 9: Heat transfer

Even the most efficient car engine only turns about 1/3 of the energy in the fuel into useful power at the wheels. The remaining 2/3 mainly comes out as heat. For example, a Formula 1 car produces 800hp (about 600kW) at the wheels but makes twice this much heat (1200kW). A standard central heating boiler is rated at about 15kW, so this means that at full power the Formula 1 car throws out enough waste heat to keep 80 houses warm and cosy.  Since they don't have fans, and simply rely on forward motion to stuff enough cooling air through the radiators, it's little wonder that they run the risk of overheating on the starting grid.

Heat can be shifted from one place to another, from where it isn't wanted to where it is, in three different ways. Conduction, convection and radiation.

**Conduction:** Some materials conduct heat better than others but the basic idea is the same. If you heat up some of the atoms in the material they'll jiggle about more rapidly, bang into their neighbours and pass some of the heat on.  The same effect can be seen in a mosh pit. This direct transfer of heat, between neighbouring atoms, is called conduction.

**Convection:** Another way of getting heat from one place to another is to simply pick up a warm object and carry it there. For example, a typical central heating system heats up water in the boiler and then pumps the hot water to the radiators.  This sort of process is often called "forced convection" to distinguish it from "natural convection" which occurs when warm air, or liquid, expands and then rises.

**Radiation:** It's about 90 million miles from the Sun to the Earth and most of that distance is fairly empty space.  The heat gets here by radiation. Our eyes are sensitive to just a small part of the electro-magnetic spectrum. If you heat up a lump of metal it will first of all glow red, then yellow then white and, if you could get it hot enough, would begin to glow blue.  You and I, on the other hand, are gently pumping out infra red radiation.

Perhaps surprisingly, it turns out that surfaces that are good at absorbing radiation (e.g. black ones) are also really good at emitting it. This is why the fins on an air cooled motor bike engine are often painted black and why a shiny kettle loses heat more slowly than a grubby one.

# Part10: Heat calculations

If you want to work out how much heat to put in to heat something up, or how much it would give out as it cools down, you need to know three things.

- What it's made of it.
- How much of it there is (it's mass).
- The change in temperature.

The **specific heat capacity** of a material is the amount of heat it takes to change the temperature of 1 kg of that material by 1K. It's units are joules per kilogram per Kelvin ( J/kg/K ). So,

**heat** (in or out) = **mass** x **specific heat capacity** x **temperature change**

To work out how much heat you'd need to melt or vaporise something, or how much heat they'd give out if they condensed or froze, you need the **Specific latent heat.** This is defined as the amount of heat needed to change the state of 1kg of the substance at its melting or boiling point without any change of temperature. So,

**Heat** (in or out) = **mass** x **specific latent heat of fusion** (melting) **or vaporisation** (boiling)

If you want to work out how quickly heat will flow through an object you need to know

- What it's made of.
- It's surface area.
- It's thickness.
- The temperature difference between the surfaces.

The **thermal conductivity** of a material is defined as the amount of heat which would flow each second between opposite faces of a 1 metre cube of the material (if you had one) when there's a temperature difference across those faces of 1K. Phew!!

It might be easier to work back from what we know.

- The larger the surface area that heat can flow through the faster it will.

- The thicker the material it has to go through the slower it will.
- The bigger the temperature difference the faster it will

This gives us the equation

**Heat per sec = (area** x **temp. diff** x **thermal conductivity) / thickness**

Put the area, temperature difference and thickness all equal to 1 and you get the definition for thermal conductivity.

The **thermal resistivity** of a material is the opposite of its thermal conductivity. It's the temperature difference you'd need to get heat to flow at the rate of 1W (1J/s) between opposite faces of a 1metre cube (again, if you had one).

The **thermal resistance** of something is the temperature difference needed to get heat to flow at 1W though each square metre. This depends on what it's made of and it's thickness.

**thermal resistance = thickness** (metres) **/ thermal conductivity**

**or                              = thickness x thermal resistivity**

If you make something from layers of different materials the thermal resistances of the layers simply add up.

Adding up thermal resistances like this, and adding a bit for surface effects, makes it easy to calculate what are known as **U-values**. These are used to work out how much heat you'll lose through your walls, roof or windows.

The rate heat is lost is given by

**heat loss per sec = area** x **U-value** x **temperature difference**

and the U-value is given by

**U-value = 1/ (sum of thermal resistances of the parts)**

or just look it up in a reputable table. It will have units of $W/m^2/K$

# Sources of Energy

## 1: Energy from the Sun

Most of the energy we use on Earth came from the Sun. The heat from the Sun not only warms us up but also powers the Earth's weather, plants use sunlight to grow and fossil fuels are just the long dead remains of things which once grew. Plus, nowadays we can use photovoltaic panels to make electricity directly from sunlight.

**Heat :** You can easily heat a house using the Sun by simply putting big windows on the side which faces the Sun and little windows on the side which doesn't and, with a bit more effort, you can use solar panels to heat your hot water. Obviously, there's no point rapping the Sun's heat if you're just going to let it wander off again through a poorly insulated loft or draughty windows.

**Weather:** The Sun heats the ground at the equator more than it does anywhere else just because it's higher in the sky. This causes warm air to rise up at the equator and sucks cooler air in from the North or South. This is what drives the Earth's weather. In addition, because the Earth spins on its axis this makes air the currents swirl as they move North or South. So, a wind turbine gets its energy from the Sun.

**Plants:** A plant's best trick is to use solar energy to combine carbon dioxide and water to form carbohydrates (cellulose and sugars etc.).In the process they give off oxygen. Us animals not only benefit from the oxygen but can also use the carbohydrates as fuel (either for our own bodies or to burn to make heat).

**Fossil fuels:** You can think of the energy in fossil fuels as trapped prehistoric sunlight. We are currently burning fossil fuels about a million times as quickly as they were originally formed.

## Where the Sun gets its energy

As far as we can tell, the Sun has been about as bright as it is now for the last 4000 million years. It was worked out in the early 1900s that if the Sun was made of coal it would have burnt out in a mere 6000 years. At the time, no one could figure out what sort of energy it could have used to burn that brightly for that long.

Nowadays we know that stars, including the Sun, get their energy from nuclear fusion. The same sort of energy that's released in an H-bomb. By working out the total energy that the sun produces, which you can do by seeing how much reaches each square metre of the Earth and then working out how much must shoot straight past the Earth into the rest of the solar system, it's quite easy to show that it's about the same as 4 billion 20MT H-bombs going off every second.

So, the Sun is a great big nuclear explosion in the sky held together by its own gravity. The next time anyone dreams about fusion power plants on Earth just point to the great big one in the sky and suggest they start making better use of the one we've already got.

## 2 Energy from the Earth

Keep digging and you'll eventually find hot rocks. In some parts of the world, Iceland, New Zealand and parts of the USA, this **geothermal energy** is close to the surface and easy to get at. In others you have to drill holes, fracture rocks and squirt down water which gets heated and comes back up hot.

The Earth was formed from a spinning cloud of dust and rocks and at first it was very hot. But, even though it's had plenty of time to cool down it doesn't show much sign of actually doing it. This means that there must be some other source of energy at work deep beneath the Earth's surface. It turns out that this comes from the radioactive decay of heavy elements like Uranium. So, the ultimate source of geothermal energy, and the power that drives earthquakes and volcanoes, is radioactive decay.

Of course, you might wonder where the radioactive elements got the energy to decay. If you did, you'd find out that whilst the stars can use nuclear fusion to make light elements, all the way up to Iron, the only way you can make the heavier elements, like lead or Uranium, is in a cataclysmic stellar explosion called a Super Nova.

In fact, it turns out that our Sun, and our solar system, was made up from material that had already been through about three of these explosions. Hence, you, me and the cat are all literally made from star dust.

# Energy in humans

Even if you're the laziest arse in the world your body still needs energy. You need it to keep your vital organs going, to repair any damage that may have occurred and to keep yourself warm. In addition, when you're young, you need it just to get bigger, for growth.

Right now, sitting on my proverbial, I'm running at about 100W. That works out at 2.4 kWh per day, 876kWh per year or 70,080 kWh over a typical 80 year lifetime.

As soon as I get up and start moving around the numbers start to go up. Typically I'll use half as much again moving around and shoving things about. So, over a typical lifetime that would be another 35,040kWh.

## The trouble with brains

Only a few animals have got brains bigger than us. At 6kg, a blue whale's brain might be 4 times bigger than mine but, at 60 tonnes, its body is 800 times bigger. Big animals need big brains because they've got more to look after. Compared to the size of our bodies, we've got by far the biggest. Even so, at about 1.5kg, my brain is only 2% of my body weight.

But what a busy 2% it is. Even when I'm just staring out of the window, the amount of visual processing going on would make the processor in this little laptop smoke. Each of my 10,000 billion neurons makes contact with 7000 others. This gives a total of about 70 million billion connections (synapses). The same size as your address book would be if you had 10 million different e-mail addresses for every person on the planet.

Keeping all these neurones charged up and running takes a lot of energy. About 20% of what I eat goes into keeping my brain going. Now, a cow standing around in a field isn't going to find 20% more food just because it can cogitate as well as ruminate. So, a bigger brain wouldn't help it survive any better than it already does. In fact it might make things worse.

Hydra are little creatures that stick to rocks in the sea and get their food by waving their tentacles in the air and catching whatever happens to drift past. They have a larval stage that swims about and has a tiny brain to help them do it. But, as soon as they settle down on the rock of their choice the first thing they do when turning into their adult form is to eat what little brain they've got.

In evolutionary terms, brains are expensive. If they weren't, then everyone would have one and then where would we be?

Paying lip service to a clever grizzly bear or something, that's where.

## The wit and wisdom of Al G Bra

I'm a virtual reality game that's played on paper.

The rules are simple, but games can get complicated.

Do the same thing to both sides of a true equation and it'll stay true.

**Rule 1**     If                  $a = b$  
                  then         $a + c = b + c$

So        If                   $y - 2 = 3$  
                  then         $y - 2 + 2 = 3 + 2$  
                  therefore      $y = 5$

**Rule 2**     If                  $a = b$  
                  then         $a - c = b - c$

So        If                   $m + 17 = q$  
                  then         $m + 17 - 17 = q - 17$  
                  therefore      $m = q - 17$

**Rule 3**     If                  $a = b$  
                  then         $a \times c = b \times c$  
                  or            $ac = bc$

So        If                   $F/3 = 6$  
                  then         $3 \times F/3 = 3 \times 6$  
                  therefore      $F = 18$

**Rule 4**     If                  $a = b$  
                  then         $a/c = b/c$

So        If                   $7a = 21$  
                  then         $7a/7 = 21/7$  
                  hence        $a = 3$

That's all the rules there are.

## Al G Bra by the rules

① If $a - 7 = 13$

then $a - 7 + 7 = 13 + 7$

∴ $a + 0 = 20$

∴ $\underline{a = 20}$

or $a - 7 = 13$

∴ $a = 13 + 7$

∴ $\underline{a = 20}$

∴ means "therefore"

② If $x + 3 = 9$

∴ $x + 3 - 3 = 9 - 3$

∴ $x + 0 = 6$

∴ $\underline{x = 6}$

or $x + 3 = 9$

∴ $x = 9 - 3$

∴ $\underline{x = 6}$

③ If $\dfrac{p}{7} = 3$

∴ $\dfrac{p}{x_1} \times x' = 3 \times 7$

∴ $\underline{p = 21}$

Using the fraction rule
(dividing top and bottom by seven)

④ If $(m + 7)\,3 = 27$

∴ $\dfrac{(m + 7)\,x'}{x_1} = \dfrac{27}{3}$

∴ $m + 7 = 9$

∴ $m + 7 - 7 = 9 - 7 \implies \underline{m = 2}$

⑤ If $F = ma$

$\therefore \dfrac{F}{m} = \dfrac{\cancel{m}\,a}{\cancel{m}}$

$\therefore a = \dfrac{F}{m}$

⑥ If $3m + 6 = 18$

$\therefore \quad 3m = 18 - 6$

$\therefore \quad 3m = 12$

$\therefore \quad m = \dfrac{12}{3} = 4$

⑦ If $2(3x + 7) = 32$

$\therefore \quad 3x + 7 = \dfrac{32}{2} = 16$

$\therefore \quad 3x = 16 - 7 = 9$

$\therefore \quad x = \dfrac{9}{3} = 3$

⑧ If $2p^2 = 50$

$\therefore \ p^2 = 50/2 = 25$

$\therefore \ p = \sqrt{25} = 5$

⑨ If $mgh = \frac{1}{2}mv^2$   find $v$ ?

$\therefore 2mgh = mv^2$

$\therefore \dfrac{2\cancel{m}gh}{\cancel{m}} = \dfrac{\cancel{m}v^2}{\cancel{m}}$

$\therefore v^2 = 2gh$

$\therefore v = \sqrt{2gh}$

Falling body.
mass $= m$
acc'n $= g$
height $= h$
velocity $= V$

loss of P.E = gain of K.E
$mgh = \frac{1}{2}mv^2$
$\therefore v = \sqrt{2gh}$

⑩ If $mgh = \frac{1}{2}mv^2$   find $h$ ?

$\therefore gh = \dfrac{v^2}{2}$   (dividing by m)

$\therefore h = \dfrac{v^2}{2g}$   (dividing by g)

From the speed the stone hits you. You can work out how far it's dropped.

# Questions, Questions

Q: George is leaning on a lampost at the corner of the street in case a certain little lady goes by. How much work does he do on the lampost?

Force

Work done = Force × distance

But the lampost doesn't move ⇒ distance = 0

∴ Work done = Force × 0 = 0

So, it doesn't matter how hard he leans, if the force doesn't move he doesn't do any work.

Q: Chas and Davva take a supermarket trolley for a walk along the old canal. If they push it two miles with a force of 50N, how much work have they done?

→ 50N

←— 2 miles —→

Work done = Force × distance

1 mile is about 1600m so 2 miles are about 3200m

∴ Work done = 50N × 3200m = 160,000 J

Q: If it takes them 1½ hours, what is the average power that they use?

$$\text{Power} = \frac{\text{Work done}}{\text{time taken}} = \frac{160000 \text{ J}}{1½ \times 60 \times 60 \text{ s}} = \frac{160000 \text{ J}}{5400 \text{ s}}$$

∴ Power = 29.6 W      Enough to power the light in a fridge.

Q: Hannibal marched his elephants over the Alps to surprise the Romans by arriving from the north. How much work is done by a 4 tonne elephant as it climbs a 1200m mountain?

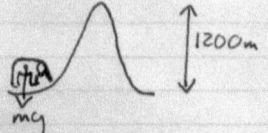
mg
1200m

As the elephant climbs the mountain it gains potential energy.

Work done = change of P.E.
= mgh

But m = 4 tonnes = 4000kg
We're on Earth so g = 10 m/s/s
and h = 1200m

$$\therefore \text{Work done} = 4000 \times 10 \times 1200$$
$$= 48,000,000 \text{ J}$$

Q: If an animal can convert 20% of the energy in its food to useful work and a Big Mac contains 540 Calories. How many Big Macs would the elephant have to eat to climb the mountain?

1 Calorie = 4200 J  $\Rightarrow$  540 Calories = 540 × 4200
$$= 2,268,000 \text{ J}$$

Now 48,000,000 J = 20% of total energy needed
= $\frac{20}{100}$ × Total Energy
$\therefore$ Total energy = $\frac{100}{20}$ × 48,000,000

$$= 240,000,000 \text{ J}$$

$\therefore$ No. of Big Macs = $\frac{240,000,000}{2,268,000}$ = 106 Big Macs

Q: Bill crimps one off with his arse 20cm (0.2m) above the water in the toilet bowl. How fast is the turd travelling when it hits the water?

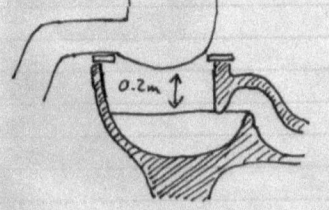

0.2m

As the turd falls it loses potential energy and gains kinetic energy

Loss of P.E = gain of K.E

$$\therefore \quad \not{m}gh = \tfrac{1}{2}\not{m}v^2$$

On Earth
g = 10 m/s/s

h = 0.2m

$$\therefore \quad 2gh = v^2$$

$$\therefore \quad v^2 = 2gh$$

$$\therefore \quad v = \sqrt{2gh}$$

$$\therefore \quad v = \sqrt{2 \times 10 \times 0.2}$$

$$\therefore \quad v = \sqrt{4} = 2 \text{ m/s}$$

On the Moon
$g_m = 1.7$ m/s/s

$$\Rightarrow \quad v_m = \sqrt{2 \times 1.7 \times 0.2} = \sqrt{0.68}$$

$$\therefore \quad v_m = 0.8 \text{ m/s}$$

On Jupiter
$g_J = 20$ m/s/s

$$\therefore \quad v_J = \sqrt{2 \times 20 \times 0.2} = \sqrt{8}$$

$$\therefore \quad v_J = 2.8 \text{ m/s}$$

# Potential Energy

Lift an object up and you'll do work against gravity. Let the object come back down again and you can get it to do work. This capacity to do work is called its potential energy.

If you're on a planet where gravity produces an acceleration $g$ then the force exerted on a mass $m$ will be given by $F = mg$.

The work done         = Force × distance
(as object falls)

$$= mg \times h$$

$$\therefore \text{Potential energy} = mgh$$

In an old pendulum clock the mechanism is often driven by a falling weight

As the weight falls a wheel turns and this powers the clock

In a hydro electric power station the potential energy of the water in the dam is converted to electrical energy by the turbines

Dam

Turbine

$h$

# Kinetic Energy

Speed something up and you'll have to do
work. This work/energy can be recovered
when it slows down again. This energy of
motion is called Kinetic Energy.

0m/s $\xrightarrow{a}$       $\rightarrow v$

$\textcircled{m} \rightarrow F$       $\textcircled{m}$

$| \longleftarrow \quad d \longrightarrow |$

A force F acts on a mass m for a time t
during which it travels a distance d

Work done to speed m up $= F \times d$

But $F = ma$ $\therefore$ work done $= ma \times d$

But $a = \dfrac{\text{change of speed}}{\text{time}} = \dfrac{v}{t}$ $\therefore t = \dfrac{v}{a}$

And average speed during all this is
half the max speed $= v/2$

During time t it travels $d = \dfrac{v}{2} \times t =$

$\therefore d = \dfrac{v}{2} \times \dfrac{v}{a} = \dfrac{v^2}{2a}$

$\therefore$ Work done $= ma \times d = m\cancel{a} \times \dfrac{v^2}{2\cancel{a}}$

$\therefore$     Kinetic energy $= \underline{\underline{\dfrac{1}{2} m v^2}}$